Arrow Jay Publishing

AI for General Contractors

Win More Bids. Run Tighter Jobs. Build a Business That Scales.

A practical guide for general contractors, small builders, and construction managers who want to use AI tools to compete smarter, protect their margins, and get more done with the team they already have.

Arrow Jay Publishing
arrowjaypublishing.com

Copyright Notice

The information in this book is provided for educational purposes only. AI tools, software platforms, and pricing mentioned throughout this book are subject to change. Always verify current pricing, features, and terms directly with vendors before making purchasing decisions.

The author and publisher make no representations or warranties with respect to the accuracy or completeness of the contents of this book. The advice and strategies contained herein may not be suitable for every situation. This book is sold with the understanding that the publisher is not engaged in rendering legal, accounting, or other professional services. For legal or other expert assistance, the services of a competent professional should be sought.

Published by Arrow Jay Publishing

arrowjaypublishing.com

ISBN (Paperback): 978-1-964326-05-4

ISBN (eBook): 978-1-964326-06-1

First Edition, 2026

Table of Contents

INTRODUCTION

How to Get the Most Out of This Book

Let me start with something honest. Most books about AI are written for people who already love technology. They are full of diagrams, acronyms, and breathless excitement about the future. This is not one of those books.

This book was written for contractors. People who wake up early, deal with problems before breakfast, and do not have time to read something that does not pay off. If that sounds like you, you are in the right place.

You do not need to understand how AI works to use it well. You do not need a background in computers, and you definitely do not need to turn your business upside down. What you need is a clear explanation of what these tools can do, which ones are worth your time, and how to plug them into the way you already work.

That is exactly what this book gives you.

Who This Book Is For

This book was written specifically for general contractors and small to mid-size builders who run between two and twenty active projects per year. It is for the person who manages bids, hires subs, deals with clients, and somehow also handles scheduling, invoicing, and safety compliance, sometimes all in the same afternoon.

If you have ever lost a bid because someone else got their numbers in faster, if you have had a client call because they had not heard from you in two weeks, or if a scope dispute cost you money that should have been yours to keep, this book addresses all of that.

How This Book Is Organized

Each chapter focuses on one problem that contractors deal with every day. You can read from front to back, which gives you the full picture, or you can jump to the chapter that solves your most urgent problem first. Either approach works.

At the end of each chapter you will find at least one ready-to-use prompt. These are designed to be used immediately, from your phone, the same day you read them. Chapter 10 is a complete prompt library with over fifty tested prompts organized by workflow. That chapter alone is worth the price of the book.

A Note on Tools and Pricing

The AI tools mentioned throughout this book reflect what was available and how they were priced at the time of writing. This is a fast-moving space. Before signing up for any paid tool, verify current pricing and features directly with the vendor. Things will have changed.

A Note on Professional Responsibility

AI tools produce drafts that require your review. They make mistakes. They do not know your local market, your specific clients, or your site conditions. Chapter 12 covers the ethics, liability, and data privacy considerations in detail. For now, the practical rule is this: everything AI produces should be reviewed by someone with the professional knowledge to confirm it is accurate and appropriate before it is used on a real project.

.

Getting the Most From Every Chapter

Each chapter in this book is designed to stand on its own. You do not have to read front to back to get value from it. If estimating is your biggest problem right now, start with Chapter 3. If client communication is keeping you up at night, go straight to Chapter 6. If you want the quick wins first, Chapter 10 is a prompt library you can use today without reading anything else. That said, if you do read from the beginning, each chapter builds on the last in a way that gives you a complete picture of how AI fits into a modern GC operation. However you approach it, the goal is the same: less time on paperwork, better documentation, more professional communications, and more time doing the work that actually grows your business.

CHAPTER 1

Why General Contractors Need AI Right Now

The construction industry has always been competitive. That is nothing new. What is new is the speed at which the gap is opening up between contractors who have started using modern AI tools and those who are still doing things the way they have always done them.

This is not about technology for technology's sake. It is about survival and growth in an industry that is under more pressure than it has been in a generation. Labor is harder to find. Materials cost more and fluctuate unpredictably. Clients expect faster responses, better documentation, and more transparent communication than ever before. And your competitors, some of them, are already using tools that make them faster, sharper, and more professional than they used to be.

This chapter lays out the reality of where the construction industry stands right now, why these pressures are not going away anytime soon, and why the contractors who start experimenting with AI in the next twelve months will have a real advantage over those who wait.

439K	**$177B**	**5-9%**	**26%**
additional workers needed in construction in 2025 (ABC)	lost annually to rework and change orders in U.S. construction	of total project cost consumed by rework on average	of rework caused directly by miscommunicatio n

The Labor Crisis Is Not Going Away

The construction industry is short on workers. Not slightly short, but significantly short, and the numbers tell a difficult story. According to the Associated Builders and Contractors, the industry needed to attract approximately 439,000 new workers in 2025 alone just to meet existing demand. A separate analysis by the National Association of Home Builders put the annual need even higher, at 723,000 new workers per year when accounting for retirements and natural attrition.

The numbers behind this shortage are structural, not temporary. Nearly one in four construction workers is now over the age of 55, meaning retirements are accelerating. Meanwhile, only three percent of young adults in one major survey said they would consider a career in the skilled trades. Trade school enrollment has dropped for years, and the pipeline of incoming workers is not keeping up with the rate at which experienced people are leaving.

The immigration workforce, which has historically made up about one in three construction trade workers, is also under pressure due to policy changes. That combination of an aging workforce, declining new entrants, and reduced immigration is creating a structural constraint on capacity that higher wages alone cannot quickly fix.

What does this mean for you as a GC? Your best workers cost more, are harder to replace when they leave, and are in demand from multiple employers simultaneously. You need to be smarter about how you deploy the people you have, and you need to reduce the amount of time everyone on your team, including you, spends on tasks that do not require skilled labor.

AI does not swing a hammer. But it can handle hours of documentation, communication, and administrative work every week, which frees your people to do what only skilled humans can do.

Margins Are Under Pressure From Multiple Directions

Construction has always operated on thin margins. But the combination of recent years has been particularly punishing. Material costs spiked dramatically during and after the pandemic and have remained volatile. Fuel costs affect every delivery and every piece of equipment on your site. And labor, as we just covered, is expensive and scarce.

The result is that the margin for error on a given project is smaller than it used to be. A missed scope item, an unexpected delay, or an estimate that was three percent off does not just feel bad. It can turn a profitable job into a break-even one or worse.

Rework is one of the biggest margin killers in the industry, and it does not get discussed as much as it should. Research from the Construction Industry Institute and Autodesk has put the cost of rework at five to nine percent of total project cost on average. When you run those numbers against a typical GC project, you are looking at a meaningful hit to your bottom line on every project where rework occurs. And rework does not have to mean a major structural mistake. It includes anything that had to be done more than once: a sub who installed the wrong material, a framing layout that did not match the updated plans, a scope that was misunderstood because the documentation was not clear enough.

The data on what causes rework is instructive. Poor collaboration is behind 48 percent of rework incidents.

Miscommunication alone accounts for 26 percent. That means more than half of all rework in construction comes from people not having the right information at the right time, or from someone misunderstanding what was communicated.

AI addresses both of those problems directly. Better documentation, clearer scope language, more consistent communication, and faster access to project information all reduce the conditions that lead to rework.

Change Orders Are Costing You More Than You Realize

Change orders are a normal part of construction. But they have become an industry-wide profitability problem that deserves its own conversation. Research across construction projects has found that change orders account for eight to fourteen percent of a typical project's total contract value. On distressed projects, that number can climb to 25 percent. The change order process itself, meaning the back and forth of drafting, submitting, waiting for approval, and sometimes disputing, contributes to ten to twenty percent of project timeline delays.

In North America in 2024, the average value of construction disputes reached 60 million dollars, up 40 percent from the prior year, with disputes taking an average of 12 and a half months to resolve. That is a statistic worth reading twice.

Most change order disputes are not about the work itself. They are about what was included in the original scope, what was excluded, and what the documentation shows. Contractors who have clear, well-written scope documents, consistent change order procedures, and thorough daily logs are in a fundamentally better position when a dispute arises than those who rely on memory, informal emails, and handshake agreements.

AI helps with all of that. It drafts scope language more precisely. It generates change order narratives that are professional and factual. It creates and organizes daily logs that provide a clear record of what happened on site and when.

Your Competitors Have Already Started

This is the part that tends to get people's attention. The contractors adopting AI right now are not mostly large companies with technology departments and multi-million dollar software budgets. Many of them are small and mid-size GCs, people running businesses very similar to yours, who found a few tools that saved them time, got comfortable with them, and then kept building.

What those contractors are experiencing is a cumulative advantage that compounds over time. When your estimating takes two days instead of five, you can respond to more opportunities. When your proposals look more professional than your competitors, you close a higher percentage. When your clients get consistent weekly updates, they call less and refer more. Each one of these improvements is modest on its own. Together, they add up to a business that feels more organized, more professional, and more capable, even if the underlying quality of your work has not changed at all.

Research from the construction technology sector has found that each additional technology a construction company implements is associated with a 1.14 percent increase in expected revenue. For a contractor doing $5 million in annual revenue, that is $57,000 per additional tool that gets genuinely integrated into workflows. The math compounds quickly.

What AI Can and Cannot Do for a GC

Before diving into the practical applications, it is worth being clear about what AI tools actually do well and where they fall short, because the contractors who get the most value from these tools are the ones who use them within their actual capabilities rather than expecting magic.

What AI does well for GCs:

- Drafting: proposals, emails, change orders, scope documents, notices, reports
- Reviewing: contracts, specs, sub bids, scope documents for gaps and risks
- Organizing: field notes into professional logs, raw data into readable formats
- Researching: code requirements, material options, industry standards
- Communicating: translating technical information into client-friendly language
- Analyzing: comparing bids, identifying inconsistencies, flagging anomalies

What AI does not do well:

- Verifying facts with certainty - it can hallucinate plausible-sounding but incorrect information
- Knowing your local market - labor rates, material costs, and sub relationships are yours to supply
- Replacing professional judgment - engineering, safety, and legal decisions require licensed professionals

- Taking action - AI produces drafts and analysis; execution is still yours

The practical rule that runs through this entire book is simple: AI produces drafts that you review. It does not replace your expertise. It amplifies it.

The Bottom Line on Why Now

There has never been a better time to start using AI tools in your construction business. The tools are more capable than they have ever been, easier to use, and more affordable. The learning curve is measured in hours, not months. The payback period on the time invested is often weeks, not years.

The contractors who are waiting for the technology to mature or for someone else to figure it out first are already behind. Not catastrophically, not yet, but the gap is growing. The ones who start now will have a real habit, a refined prompt library, and measurable results by the time AI tools become something the whole industry takes for granted.

That is the situation. The rest of this book is the practical path forward.

KEY TAKEAWAYS
Labor shortages, margin pressure, and client expectations are structural challenges that are not going away.
AI does not solve these problems directly, but it reduces the time and effort required to manage documentation, communication, and estimation.
The contractors getting the most from AI are small and mid-size GCs, not large corporations.
The time to start is now - before the gap between early adopters and everyone else becomes a competitive disadvantage.

CHAPTER 2

AI Basics for Busy Contractors

You do not need to understand how AI works at a technical level to use it effectively. This chapter gives you the practical foundation you need: what the main tools are, how to choose between them, how to use them from your phone, and how to write prompts that get useful results.

Think of this chapter as orientation. If you are already comfortable with AI tools, you can skim it and move on. If you are newer to these tools, reading this chapter carefully will make everything that follows more useful.

How AI Language Tools Actually Work

The AI tools we will be using throughout this book are called large language models, or LLMs. They have been trained on enormous amounts of text, enough to develop a detailed understanding of how language works across thousands of subjects, including construction.

When you type a question or a request, the AI predicts the most useful response based on the patterns in its training data. It is not looking up answers in a database, and it is not thinking the way a person thinks. It is generating language that fits the pattern of what a helpful, knowledgeable response would look like based on everything it has learned.

This is why AI tools are genuinely useful for writing, reviewing, and analyzing documents. They have processed millions of examples of proposals, contracts, emails, and reports, and they can generate new content that follows the patterns of good professional writing.

It is also why they occasionally get things wrong. Because they generate responses based on patterns rather than looking up verified facts, they will occasionally state something that is incorrect with complete confidence. This is called hallucination in the AI world, and it is a real limitation. It is not a reason to avoid these tools, but it is a reason to always review what they produce before using it.

The Big Three: ChatGPT, Claude, and Gemini

Three general-purpose AI tools dominate the market and are worth knowing about: ChatGPT from OpenAI, Claude from Anthropic, and Gemini from Google. Grok from xAI is also a growing option. All four are capable enough to handle everything described in this book. The differences between them are real but not dramatic enough to matter much when you are starting out.

ChatGPT is the most widely used AI tool globally. It has the largest ecosystem of integrations, the most tutorials and community resources, and a very capable free tier. The paid version, ChatGPT Plus at $20 per month, gives you access to their most powerful models and image generation capabilities. For most contractors, ChatGPT is a natural starting point because of how much help is available online when you get stuck.

Claude from Anthropic tends to produce longer, more detailed responses and is particularly strong at handling long documents. If you need to paste in a full contract or a lengthy spec section and ask questions about it, Claude often handles this more effectively than the alternatives. Claude also tends to follow complex, multi-part instructions more reliably, which makes it useful for detailed prompt templates like the ones in Chapter 10.

Gemini from Google integrates directly with Gmail, Google Docs, and Google Drive. If you are already using Google Workspace for your business, Gemini can work inside the tools you are already using rather than requiring you to switch to a separate app. For contractors who manage most of their communication and documentation through Google, this integration advantage is significant.

Microsoft Copilot integrates with Word, Excel, Outlook, and Teams. If your business runs on Microsoft 365, Copilot adds AI capabilities to the tools you already use without requiring separate subscriptions.

For starting out, the recommendation is simple: if you use Google Workspace, start with Gemini. If you use Microsoft 365, start with Copilot. If you are not committed to either ecosystem, start with the free version of ChatGPT or Claude and upgrade to a paid tier after you have confirmed it is saving you time.

Tool	Best For	Free Tier	Paid
ChatGPT	Broad use, large ecosystem	Yes	$20/mo (Plus)
Claude	Long documents, complex prompts	Yes	$20/mo (Pro)
Gemini	Google Workspace users	Yes	$19.99/mo

Tool	Best For	Free Tier	Paid
Copilot	Microsoft 365 users	Limited	$30/user/mo
Grok	X/Twitter integration, real-time data	Limited	Included with X Premium+

Specialized Construction AI Tools

Beyond the general-purpose tools, a growing category of AI tools has been built specifically for construction. These tools are trained on construction data, designed for construction workflows, and in some cases integrate directly with your existing project management software.

Estimating and takeoff tools like Togal.AI and STACK use AI to extract quantities from plan drawings, dramatically reducing the time required for manual takeoff. These tools are covered in detail in Chapter 3.

Project management platforms like Procore and Autodesk Construction Cloud are adding AI capabilities for RFI automation, schedule analysis, and document processing. If you are already on one of these platforms, the AI features may be available to you at no additional cost.

Field documentation tools like CompanyCam and OpenSpace use AI for photo organization, progress tracking, and report generation. These are covered in Chapter 8.

Specialized contract review tools like Document Crunch are trained specifically on construction contracts and can flag non-standard clauses faster than general-purpose AI tools. Covered in Chapter 11.

The general advice on specialized tools: start with general-purpose AI and move to specialized tools when you have identified a specific workflow where the general tool is not meeting your needs. Most contractors find that general-purpose tools handle 80 percent of their AI use cases, and specialized tools are worth the additional cost only for workflows where they do genuinely better.

Phone-First: Building an AI Habit That Fits Your Schedule

One of the most important mindset shifts for contractors is treating AI as a phone tool, not a desk tool. Construction businesses do not run at a desk. They run on job sites, in trucks, in client meetings, and in brief gaps between problems. If AI is something you only use when you are at your computer, you will not build the consistent habit that makes it genuinely valuable.

All the major AI tools have mobile apps that work well. The workflow that most field contractors find most sustainable is:

- Open the app on your phone.
- Think about the two or three things you need to communicate that morning.
- Dictate or type a brief description of each item and ask the AI to draft it.
- Read each draft, adjust anything that does not sound right, and send.

This approach takes ten to twenty minutes for most contractors and replaces thirty to sixty minutes of writing time. More importantly, it produces more professional-sounding communications than most contractors produce when writing under time pressure at the end of a long day.

Voice dictation works well with AI tools. You do not need to speak in perfect sentences or organize your thoughts before you start. Describe the situation conversationally, and ask the AI to turn it into a professional communication. The AI will handle the structure and the language.

How to Write Better Prompts

The quality of what AI produces is directly related to the quality of your instructions. A vague prompt produces a generic response. A specific, well-structured prompt produces something you can actually use with minimal editing.

The most important thing to understand about prompting is that AI tools respond well to context. The more you tell the AI about your situation, your goals, and what you want the output to look like, the better the output will be.

The four elements of a good prompt:

- **Role:** Tell the AI what kind of expert to be. "You are an experienced general contractor." This shapes the perspective and vocabulary of the response.
- **Context:** Describe your specific situation. What is the project type? Who is the client? What is the issue? The more specific, the better.
- **Task:** Tell the AI exactly what you want it to produce. A draft email, a list of questions, a contract review, a schedule.
- **Format:** Tell the AI how you want the output structured. Three paragraphs, a bulleted list, a formal letter with a subject line.

Weak prompt: "Write me an email about a delay."

Strong prompt: "You are an experienced general contractor. Write a professional email to a residential client explaining that their kitchen remodel is running five days behind schedule due to a cabinet delivery delay from the supplier. The client is detail-oriented and appreciates direct communication. Acknowledge the delay clearly, explain the cause briefly, tell them the new expected completion date, and confirm whether any decisions are needed from them. Keep it to two short paragraphs. Professional but warm tone."

The strong prompt takes thirty seconds longer to write but produces something you can send with minimal editing. The weak prompt produces something generic that requires significant rewriting.

Over time, you will develop a set of prompt templates for your most common situations. Chapter 10 gives you a complete library of tested prompts for construction scenarios. Start with those and modify them to fit your specific situation.

What AI Is Good At and Where It Falls Short

Understanding where AI helps and where it does not will save you time and prevent frustration. Here is an honest assessment based on how contractors actually use these tools.

Where AI consistently delivers value for GCs:

- Turning bullet points or field notes into professional communications
- Reviewing documents for potential issues, gaps, or unusual language
- Generating first drafts of standard documents like proposals, change orders, and scope documents

- Summarizing long documents so you can identify what needs your attention
- Answering general research questions about codes, standards, and industry practices

Where AI requires extra caution:

- Specific local code requirements - always verify against your local authority having jurisdiction
- Current pricing information - AI training data has cutoffs and may not reflect current market rates
- Legal interpretations - AI can identify issues but cannot give legal advice
- Specific product specifications - verify technical details directly with manufacturers

The contractors who use AI most effectively are not the ones who trust it blindly. They are the ones who understand its strengths and use it accordingly, while bringing their own professional knowledge to review and validate the outputs.

A Note on Getting Your Team On Board

If you are a solo operator or a very small GC, adoption is straightforward: you decide to use the tools, and you use them. For GCs with staff, the adoption question is more complex.

The most common mistake is mandating adoption without demonstrating value first. People do not change how they work because they are told to. They change when they see something clearly making their job easier.

The approach that works: use AI yourself for a month, get comfortable with it, identify two or three workflows where it consistently saves you time, and then show those specific workflows to one or two people on your team. Walk them through the prompt, show them the output, and let them see how much time it saves. Let that demonstration do the persuading rather than the mandate.

Team adoption is covered in more detail in Chapter 13, along with a full 90-day implementation plan. For now, the starting point is you. Build the habit yourself before worrying about anyone else.

PROMPT TO TRY

You are a general contractor. I need to send a quick email to my plumbing subcontractor reminding them that their work on the Oak Street project needs to be complete by Friday because the drywall sub is starting Monday. The plumber tends to be responsive but occasionally needs a nudge on deadlines. Keep the email professional but direct. One paragraph.

CHAPTER 3

Winning More Bids with AI Estimating

Estimating is where most GCs spend the most time and where the consequences of errors are the most expensive. Getting your numbers wrong in either direction costs you: too high and you lose the bid, too low and you lose money on the job. Getting your numbers in on time is also critical, because late bids often do not get considered at all, regardless of how good the price is.

AI does not replace your estimating expertise. It does not know your local labor rates, your sub relationships, or the idiosyncrasies of your specific market. What AI does is help you move faster through the parts of the estimating process that are time-consuming but do not require your professional judgment, so you can spend more of your time on the parts that do.

This chapter covers how to use AI tools to speed up the estimating workflow, improve the quality of your scope analysis, and build a historical cost database that makes your estimates more accurate over time.

The Real Cost of Slow Estimating

Most GCs underestimate how much slow estimating costs them, not just in time but in missed opportunities. When it takes your team five days to put together a bid that a competitor can assemble in two, you are not just losing time. You are limiting how many opportunities you can evaluate in a given month, which limits how selective you can be about which jobs you pursue.

The contractors who can respond to more bid opportunities with quality estimates have a structural advantage: they can be selective about which ones they actually submit, they can pursue more projects in high-margin categories, and they are less dependent on any single bid winning.

AI does not magically cut your estimating time in half on the first day. But contractors who have built AI into their estimating workflow consistently report 30 to 50 percent reductions in time spent on preliminary takeoff and scope review. For a GC doing ten bids per month, that is a significant recovery of capacity.

How AI Changes the Estimating Workflow

The traditional estimating workflow for a GC looks something like this: receive bid package, review plans and specs manually, do takeoff by hand or with a basic software tool, assemble costs from memory and past projects, write up a proposal, submit. At every stage, you are starting from scratch.

An AI-assisted workflow looks different. Here is a practical breakdown of how AI fits into each stage of a typical commercial bid:

Stage 1: Initial Scope Review

Before you spend hours on takeoff, you need to understand the exact scope of what you are bidding. AI tools can help you do this initial review faster and more thoroughly than a quick skim of the plans.

Upload the specifications or paste key sections into an AI tool and ask it to summarize: what is included, what is excluded, what are the key performance requirements, and what are the unusual or non-standard items. Ask it to flag anything that is ambiguous or that could be interpreted differently by different contractors.

This analysis takes fifteen to twenty minutes with AI assistance instead of two to three hours of careful manual reading. It also tends to catch things that a quick skim misses, because the AI does not skip sections when it is tired or distracted.

Stage 2: AI-Assisted Takeoff

For projects where digital plans are available, specialized AI takeoff tools like Togal.AI or STACK can extract quantities directly from PDF drawings. These tools use computer vision to identify walls, floors, openings, and other measurable elements and calculate quantities automatically.

The accuracy of these tools varies by project type and plan quality. For straightforward commercial tenant improvements or residential projects with clean drawings, accuracy is high. For complex structural or MEP-heavy projects with multiple plan revisions, manual review of AI takeoff is essential.

The workflow is not "let AI do the takeoff and trust it." It is "let AI do the takeoff first, then use your expertise to review and correct." This review takes less time than starting from scratch, and experienced estimators catch errors quickly because they have intuition for what quantities should look like for a given project type.

Stage 3: Scope Gap Analysis

One of the highest-value applications of AI in estimating is scope gap detection. After you have assembled your preliminary takeoff and scope, paste your scope summary and each sub's bid into Claude or ChatGPT and ask it to identify what is in the project scope that is not covered by any sub bid, and what is covered by multiple sub bids with potential for double-counting.

Scope gaps are a major source of change orders and disputes. Catching them before bid day rather than after job start protects your margin and your client relationships.

Stage 4: Historical Cost Integration

AI cannot tell you what concrete costs in your market right now. But you can train it to use your historical data. Keep a running record of your actual costs by trade and project type, and feed this into your AI prompts when you are estimating similar projects. "On our last three office TI projects in this price range, we ran about $X per SF for framing. Does my current framing estimate look consistent with that range?"

This kind of sanity-check prompting is fast and valuable. It catches estimates that are significantly off from your experience before you submit them.

Stage 5: Proposal Generation

Once your estimate is assembled, AI dramatically speeds up proposal writing. Feed your bid numbers, project description, client information, and any relevant notes into the AI and ask it to draft a professional proposal following your standard format. The draft will need editing, but it gives you a solid starting point in minutes rather than hours.

Here is what a complete AI-assisted bid workflow looks like in practice:

- Receive the bid package including plans, specifications, and bid form.
- Upload plans to Togal.AI or STACK for automated takeoff. Review and correct quantities.

- Paste the spec sections into Claude and ask for a summary of key requirements, unusual items, and potential scope ambiguities.
- Ask the AI to run a scope gap analysis on the full project scope and each trade.
- Assemble your costs using the takeoff quantities and your cost data.
- Ask the AI to review your assembled bid for any line items that seem significantly out of range for this project type, flagging anything more than 20 percent above or below typical ranges you describe.
- Have the AI draft your proposal using the structured bid as input.

Building Your Own Historical Cost Database

One of the most valuable long-term investments you can make in your estimating practice is building a structured database of your actual project costs. Not published cost guides, but your costs, for your project types, in your market.

AI makes this easier to build and easier to use. After each project closes, spend thirty minutes recording the actual costs by trade division, the project type and size, the location, and any unusual factors that affected costs. Store this in a simple spreadsheet or document. Over time, this becomes a searchable reference that you can paste into AI prompts when estimating similar projects.

The more projects you add to this database, the more valuable it becomes. Two years of consistent tracking gives you a reference that is more accurate for your specific business than any published cost guide, because it reflects your actual sub relationships, your market, and your operational overhead.

What to Do When Plans Are Not Available

Many estimating situations do not involve complete design documents. Residential remodelers often bid from a conversation and a site visit. Early-stage commercial projects may have schematic drawings only. Design-build GCs may need to develop preliminary costs before any plans exist.

AI is particularly useful in these situations because it can help you develop a comprehensive scope from a description, generate typical cost ranges for project types by size and location, identify the line items you need to include that might not be obvious from the initial description, and flag the missing information that will most significantly affect cost accuracy.

The prompt for this situation is straightforward: describe the project as specifically as you can in terms of type, size, location, and scope intent, and ask the AI to generate a preliminary budget framework with the main cost categories and typical ranges for a project of this type.

Practical Example: A Commercial Tenant Improvement Bid

Here is how a GC might use AI on a real bid for a 4,500 SF office tenant improvement in a Class B office building. The bid package includes architectural plans, MEP drawings, and a 60-page specification.

Step 1: Spec Summary (15 minutes)

Paste the key specification sections into Claude and ask: "Summarize the main requirements of this specification. Flag any unusual or non-standard requirements, any performance specifications that will require specific sub qualifications, and any items that are ambiguous about who is responsible."

The output flags that the spec requires specific LEED documentation for certain materials, that the acoustical tile specification references a non-standard manufacturer, and that the electrical spec is ambiguous about whether the GC or owner is responsible for panel upgrades.

Step 2: Scope Gap Check (20 minutes)

After receiving sub-contractor bids, paste the project scope and each sub's bid into the AI and ask for a gap analysis. The AI identifies that testing and balancing of the HVAC system appears in the mechanical spec but is not explicitly included in any mechanical sub's bid. This item is worth $8,000 to $15,000, and discovering it before award rather than after start saves a significant conflict.

Step 3: Estimate Review (10 minutes)

Paste the assembled estimate and ask: "Review this estimate for a 4,500 SF office TI in a Class B building. Are there any line items that seem significantly out of range for this project type? My direct costs total $285,000 before overhead and profit." The AI flags that the drywall line item at $85,000 seems high for this square footage at roughly $19 per SF, compared to typical ranges of $12 to $15 for standard office work, and suggests reviewing whether it includes both metal framing and drywall.

Step 4: Proposal Draft (20 minutes)

Using the bid numbers and project notes, the AI drafts a professional two-page proposal in 90 seconds. The estimator spends 20 minutes editing it to match the company voice, add project-specific details, and verify accuracy. Total proposal writing time: 20 minutes instead of 90.

Total AI-assisted time savings on this bid: approximately 3 to 4 hours. Multiply that across 10 bids per month and you are recovering 30 to 40 hours of capacity every month.

PROMPT TO TRY
Review the following project scope for a [project type and size] and give me: (1) a list of items that are ambiguous or could be interpreted differently by different contractors, (2) items commonly included on this type of project that are not explicitly addressed in this scope, and (3) items I should explicitly exclude from my bid to prevent scope creep. Format each as a separate numbered list.
[Paste scope of work here]

CHAPTER 4

Proposals, Contracts, and Change Orders

The documents you produce define your professional reputation as much as the work you build. A clear, well-structured proposal tells a potential client that you run an organized business. A precise scope of work document reduces disputes before they start. A professional change order narrative makes approval easier. AI makes all of these better, faster, and more consistent.

This chapter covers the full document workflow for a GC: from the proposal that wins the job to the change orders that protect your margin during it.

Why Proposals Matter More Than Most Contractors Think

Most contractors think of a proposal as a formality that comes after the real work of estimating. The numbers are what matter, they figure, and the proposal is just wrapping paper.

This is wrong, and it costs money.

Clients, especially for projects above $50,000, are making a decision about who to trust with their building, their money, and their time. The proposal is often the most substantive piece of evidence they have to evaluate that trust before a project starts. A proposal that is organized, specific, and professionally written signals that your company operates that way. A proposal that is vague, full of boilerplate, or difficult to read signals the opposite.

In competitive bidding situations, proposals have won jobs for contractors who were not the low bid. The client chose them because their proposal demonstrated a deeper understanding of the project, a clearer explanation of their approach, and a more professional presentation than the lower-priced alternative.

AI makes it practical to write this quality of proposal consistently, even under time pressure. The same contractor who used to dash off a two-paragraph proposal the night before bid day can now produce a structured, professional document in the same amount of time or less.

The AI-Assisted Proposal Writing Process

Here is a practical workflow for using AI to produce a professional proposal:

Step 1: Brief the AI. Before asking the AI to write anything, give it context. What is the project type and size? Who is the client, and what do you know about what matters to them? What is your competitive position, are you the incumbent, a new bidder, or one of three? What aspects of your company are most relevant to this project?

Step 2: Supply your bullet points. Do not ask the AI to invent content. Tell it what you want to include: your scope, your team, your timeline, any relevant past projects, and your pricing approach. The AI turns your bullet points into professional prose. The content stays yours.

Step 3: Specify the format. Tell the AI exactly how you want the proposal structured. The standard contractor proposal format in the prompt below works well for most commercial and high-end residential projects.

Step 4: Review and personalize. The AI draft will be professional but generic. Your job is to add the specific details that make it yours: the project manager's name, a reference to something specific the client said, the particular aspect of the project that your company is especially suited to.

Step 5: Check the numbers. AI does not verify math. Before the proposal goes out, confirm that every number in it, every date, every square footage, every allowance, matches your actual estimate.

PROMPT TO TRY
Write a professional project proposal for a [client type] hiring us for [project description]. Organize it with these sections: Project Understanding (demonstrate we grasp their goals), Our Approach (how we will execute), Scope of Work (specific inclusions), Exclusions (what is not included), Timeline (key milestones and completion date), Our Team (brief relevant credentials), and Investment (how they authorize the work). Two pages maximum. The tone should be professional but approachable, not overly formal.
[Add your bullet points about the project here]

Key Elements of a Strong Proposal

Regardless of project type or size, the proposals that consistently win share certain characteristics. Understanding these helps you brief the AI effectively and review the draft with the right eye.

Project understanding: The opening section should demonstrate that you understand what the client is actually trying to accomplish, not just the technical scope of the work. A retail client wants to open on time for a lease date. A homeowner wants to minimize disruption to their family. A corporate client wants the project to reflect well on their brand. The proposal that shows you grasp this wins more often than the one that jumps straight to scope.

Specific scope language: Vague scope is a setup for disputes. "All interior finishes as per drawings" is not a scope statement. "Drywall and tape throughout, Level 4 finish in public areas, Level 3 in back-of-house, painted in owner-selected colors per paint allowance" is a scope statement. AI can help you translate your rough scope notes into specific, unambiguous language.

Clear exclusions: What you are not doing is as important as what you are doing. Explicit exclusions prevent the "I thought that was included" conversation that becomes a change order dispute. List them clearly.

Allowances called out: If you have priced any items as allowances rather than firm bids, those allowances should be listed explicitly with the dollar amount and a brief description of what the allowance covers. Hidden allowances that get discovered later damage trust.

Professional presentation: Format matters. A proposal that is easy to read, clearly organized, and free of typos signals professionalism. AI helps with all of this, but your review catches what the AI misses.

Using AI for Contract Review

Contract review is something most GCs should do and most do not do it thoroughly enough. By the time a contract hits your inbox, you are usually already mentally committed to the project and ready to move. The contract feels like an administrative formality, and reading it carefully feels like a delay.

This is one of the most expensive mistakes in the construction business. Contracts contain provisions that can dramatically affect your rights and obligations, your ability to get paid, and your exposure if something goes wrong. Non-standard provisions that favor the owner, aggressive indemnification language, payment terms that create cash flow problems, and termination clauses that give the owner extraordinary rights are all common in contracts that contractors sign without fully understanding.

AI cannot give you legal advice and is not a replacement for an attorney. What it can do is help you read and understand contracts faster, flag the provisions that are unusual or potentially problematic, and generate a list of questions to raise with the owner or your attorney before you sign.

The workflow: paste the contract or the key provisions into Claude and ask it to review for specific risk categories. The prompt below gives you a starting point. Focus the AI on the provisions that most commonly cause problems for GCs: payment timing, change order procedures, retainage, indemnification, and dispute resolution.

For contracts above your normal project size, or contracts for clients you have not worked with before, the investment in an attorney's review is usually worth it. AI helps you prepare for that conversation by identifying the issues in advance.

PROMPT TO TRY

Review the following contract clauses for a construction project. Flag anything that is: (1) outside standard industry practice, (2) unusually one-sided in favor of the owner, or (3) likely to create disputes during project execution. For each flagged item, explain what the concern is and what alternative language I might propose. Focus on: payment timing, retainage, change order procedures, indemnification, consequential damages, termination rights, and lien provisions.

[Paste contract sections]

Writing Scope of Work Documents That Prevent Disputes

Scope disputes are the leading cause of change orders and litigation in construction. The project finishes, and the client was expecting something that you did not include, or the sub thought something was in their scope that you thought was in yours. These disputes are frustrating, time-consuming, and expensive, and the majority of them are preventable with better scope documentation at the start.

AI helps write scope documents that are specific enough to prevent most of these situations. The key is providing the AI with your scope notes, telling it to make the language explicit about inclusions and exclusions, and then reviewing the output with the question: "Could a contractor reading this reasonably interpret it differently than I intend?"

A good scope document for a subcontract should include:

- Specific work included, by location and specification reference where applicable

- Explicit exclusions that might otherwise be assumed to be included
- Interface responsibilities - who is responsible for work at the boundary between trades
- Access, sequencing, and coordination requirements
- Quality standards and finish requirements
- Any work that has been priced as an allowance rather than a firm bid

Change Orders: Writing Them So They Actually Get Approved

Change orders that get disputed or rejected often have a common problem: they are written from the contractor's perspective about the cost, rather than from the client's perspective about why the cost is justified.

A change order narrative that says "Additional work $4,200" gets questioned. A change order that says "This work was not included in the original scope because the structural drawings issued at bid did not reflect the field condition that was discovered when the existing slab was opened. The original specification called for X; the actual condition required Y. The additional cost reflects the labor and materials required to address this unanticipated condition" gives the client the information they need to understand and approve the change.

AI is excellent at writing change order narratives. Give it the facts: what the original scope said, what the actual condition was, why the additional work was required, and what it cost. The AI will produce a professional, factual narrative that makes the case without sounding defensive or adversarial.

PROMPT TO TRY

Write a professional change order narrative. Original scope did not include: [describe]. Reason this work is now required: [explain]. Additional work involved: [describe]. Cost impact: [amount]. Schedule impact: [days]. Write this as a factual business communication that makes the case for approval clearly and professionally. Do not be defensive. Reference the original scope language where relevant.

Tracking Changes and Managing the Paper Trail

One of the most valuable things AI can help with in change order management is not writing the change orders themselves, but helping you maintain the documentation trail that supports them.

Daily logs, photo captions, and email summaries that consistently document site conditions, decisions made, and instructions given create a contemporaneous record that is invaluable when a dispute arises months later. An AI-assisted documentation habit, ten minutes at the end of each day, can generate a project record that would take hours to create manually.

The practical approach: at the end of each workday, dictate or type your notes about what happened on site. Send them to the AI with the prompt: "Convert these field notes into a professional daily log entry, organized by activity, with a clear record of any decisions made, instructions given, or changed conditions encountered." Save the output. Over the course of a project, this daily log becomes your most powerful asset if a dispute arises.

KEY TAKEAWAYS
Professional proposals win jobs even when you are not the low bidder.
AI-assisted proposals are more organized and specific than time-pressured manual writing.
Contract review with AI helps you identify non-standard provisions before you sign.
Change order narratives should explain why the cost is justified, not just state the amount.
Daily AI-assisted documentation creates a project record that protects you in disputes.

CHAPTER 5

Scheduling, Timelines, and Delay Prevention

Construction schedules fail for predictable reasons. Understanding those reasons is the first step toward building schedules that hold.

Projects get behind for a small number of recurring causes: materials that do not arrive when expected, trades that are not ready when the previous trade finishes, inspections that are not scheduled far enough in advance, scope changes that are not reflected in the schedule, and weather. Of these, the first four are largely preventable with better planning and communication. AI helps with all of them.

Why Construction Schedules Fall Apart

The most common scheduling failure is not that the schedule is wrong from the start. It is that the schedule is not updated, communicated, and managed as conditions change. A schedule that sits in a drawer after kickoff is not a planning tool, it is a historical document.

The contractors who keep projects on schedule do three things consistently: they maintain a current schedule that reflects what is actually happening, they look ahead two to three weeks continuously and identify problems before they hit, and they communicate schedule status to all affected parties regularly enough that problems get surfaced while there is still time to solve them.

AI helps with all three of these. It does not replace scheduling software for complex projects, but it handles a significant part of the communication and analysis work that makes schedules useful in practice.

Sometimes it is the weather. Sometimes it is a material delivery that a supplier pushed back without telling you. Sometimes a preceding task is not done, and the next trade shows up anyway. In all of these cases, the difference between a recoverable delay and a cascading problem is how quickly you identify it and how clearly you communicate it to everyone affected.

Building a Project Schedule with AI Assistance

For most GCs, the practical use of AI in scheduling is not building the master schedule from scratch, it is building the framework, checking it for logic errors, and generating the communications that keep everyone aligned.

Framework generation: For project types you do similar work repeatedly, ask the AI to generate a schedule framework including standard phases, typical durations, and common dependencies. Use this as a starting point and adjust for your specific project conditions. This takes fifteen minutes instead of two hours.

Logic checking: After you have built your schedule, describe it to the AI and ask it to identify potential sequencing problems, missing dependencies, or phases where the duration seems short for the scope described. AI is good at catching logical errors that are easy to miss when you are too close to the project.

Long-lead identification: Ask the AI to review your project scope and identify materials or equipment that typically require long lead times and should be ordered early. For MEP equipment, specialty finishes, and custom fabrication, early identification of long-lead items prevents delays that feel sudden but are entirely predictable.

PROMPT TO TRY
Generate a project schedule for the following project. Include logical sequencing with dependencies noted, estimated durations for each phase, key milestones, long-lead items that require early ordering or coordination, and any activities where sequencing problems commonly occur.
Project type: [describe]
Trades involved: [list]
Start date: [date]
Required completion: [date]
Known constraints: [phasing, occupancy, inspections, owner milestones]

The Two-Week Look-Ahead: Your Most Practical Scheduling Tool

The two-week look-ahead is the most practical scheduling tool for most GCs. It bridges the gap between the master schedule and daily work. It is specific enough to be actionable, short enough to be current, and long enough to allow you to solve problems before they arrive.

The challenge with two-week look-aheads is generating them consistently. When you are busy, updating the look-ahead feels like one more thing to do. AI removes most of the friction.

Give the AI your current project status, what was completed last week, what is planned for the next two weeks, what deliveries are expected, and what inspections are scheduled. The AI formats this into a professional look-ahead document in two to three minutes. You review it, make any corrections, and distribute it to your team and subs. The whole process takes ten to fifteen minutes instead of forty-five.

Specialized Scheduling Tools Worth Knowing

For GCs managing complex projects with multiple concurrent activities and tight sequencing, general-purpose AI tools have limitations. Specialized scheduling tools with AI capabilities are worth considering.

ALICE Technologies uses AI to model different resource and sequencing scenarios, allowing you to simulate the impact of schedule decisions on cost and completion before you commit. Best for complex commercial or industrial projects where optimization of sequencing has significant cost implications.

Planera focuses on pull planning and AI-assisted schedule intelligence, with tools for multi-trade coordination and collaborative planning sessions. Good fit for GCs who use lean construction principles.

SmartPM specializes in schedule analytics and delay analysis. If you are on projects where schedule disputes and delay claims are a real risk, SmartPM's AI-powered analysis tools can help you document schedule performance and build the evidence base for delay claims or defenses.

Communicating Delays Professionally

When a delay occurs, how you communicate it matters almost as much as how you address it. Clients who receive a clear, factual, early notice of a delay with a specific explanation and a recovery plan are far more forgiving than clients who find out a project is behind through their own observation.

AI is excellent at drafting delay notices. Give it the facts: what the delay is, when it was identified, what caused it, what the schedule impact is, and what you are doing about it. The AI produces a professional notice that gives the client what they need to understand the situation without sounding defensive or making admissions that could create legal problems.

Recovery Scheduling: Getting Back On Track

When a project falls behind, the natural response is to add overtime and weekend work. This is sometimes the right answer, but it is rarely the only option, and it is often not the most cost-effective one.

Before committing to a recovery approach, ask the AI to help you think through the options. Give it the specifics: how far behind you are, what resources are available for recovery, and what the constraints are in terms of budget and site conditions. Ask it to generate multiple recovery scenarios with different cost and time tradeoffs.

This takes fifteen minutes and almost always surfaces at least one alternative that is worth considering before you default to the most expensive approach.

KEY TAKEAWAYS
Two-week look-aheads are the most practical scheduling tool for most GCs.
AI can generate schedule frameworks and check for logic errors in your sequences.
Early identification of long-lead items prevents delays that feel sudden but are predictable.
Delay notices should be factual, early, and specific about cause and recovery plan.

CHAPTER 6

Client Communication and Relationship Management

Client communication is where many GCs leave significant money on the table, not because they do bad work, but because their clients do not know what is happening. The client who does not get regular updates starts to wonder whether things are going well. The wondering becomes worry. The worry becomes phone calls. The phone calls interrupt your day and create a dynamic where you are constantly reactive rather than proactive.

The contractors who get the most referrals and the most repeat business are not always the ones who do the best technical work. They are the ones who make their clients feel informed, respected, and confident throughout the project. AI makes it practical to be that contractor consistently, even when you are managing multiple projects and a full schedule.

The Communication Gap That Costs You Referrals

Research on what drives client satisfaction and referrals in construction consistently identifies communication as the top factor, ahead of quality of work and on-time completion. Clients expect quality, and when they get it, they consider it the baseline. What differentiates their experience is how they felt during the project.

The clients who feel informed and respected become your most reliable source of referrals. Your business will grow primarily through referrals, and referrals come from satisfied clients. Satisfied clients will refer you to new business. They hire you for their next project. They tell their friends. They leave reviews that bring in new clients you have never met.

The investment required to generate these outcomes is not large. For most GCs, a consistent weekly update system, a professional response to every client inquiry within a business day, and a thoughtful closeout communication is enough to differentiate the client experience significantly.

The Weekly Update System

The most impactful change most GCs can make to their client communication is implementing a consistent weekly project update. Not an occasional email when something significant happens, not a call when the client reaches out first, but a proactive, structured update at the same time every week for every active project.

Here is the system that works: Every Friday morning, spend five minutes per project dictating your status notes into your phone. What was completed this week. What is planned for next week. Any decisions the client needs to make. Any items affecting the schedule or budget. Send the transcription to your AI tool with the prompt below and ask it to produce a professional client update email. Review it, make any necessary adjustments, and send.

For a GC with five active projects, this entire process takes thirty to forty-five minutes. The result is five professional update emails that reduce client anxiety, prevent mid-project phone calls, and build the reputation for communication that drives referrals.

PROMPT TO TRY
Write a weekly project update email for a client. Three short paragraphs maximum. Professional but warm tone.
Project: [brief description]
Completed this week: [bullet points]
Planned next week: [bullet points]
Decisions needed from client: [list or "none at this time"]
Items affecting schedule or budget: [list or "none"]
Milestone or good news to highlight: [optional]

Handling Difficult Conversations

The most uncomfortable communication situations in construction are the ones that cannot be avoided: telling a client their project is behind schedule, explaining that a cost has increased, addressing a quality concern, or delivering news that affects their move-in date.

These conversations go better when they are structured clearly and delivered proactively. The client who hears difficult news promptly, with a clear explanation and a specific plan for addressing it, responds very differently than the client who finds out the same news by accident or after a delay.

AI is useful for preparing for these conversations in two ways: drafting the written communication, and helping you think through what you want to say before a call or in-person meeting.

For written communication about difficult news, give the AI the facts and ask it to structure the message clearly. Use this structure:

- State the situation clearly and factually, without minimizing or dramatizing.
- Explain the cause to the extent you know it, without deflecting responsibility or assigning blame prematurely.
- Describe what you are doing about it, specifically and concretely.
- Tell the client what they need to know or do, and what happens next.

This structure works in emails, in letters, and as a framework for a conversation. It gives the client the information they need without creating unnecessary alarm or legal risk.

Proposal Follow-Up: Closing More of What You Bid

Most GCs submit proposals and then wait. Some follow up once. Very few follow up in a way that adds value rather than just asking "have you made a decision yet?"

AI helps you write follow-up communications that add value and keep you top of mind without being pushy. The most effective follow-up references something specific about the client's project or goals, adds one piece of relevant information or perspective, and gently invites a next conversation without pressure.

This kind of follow-up is easy to write when you are not in a hurry. It is nearly impossible to write well when you are busy. AI makes the quality version possible even on a tight schedule.

Client Relationship Management Without a CRM

For small GCs, a full CRM system is often overkill. But some basic tracking of your client relationships, past projects, and communication history is valuable, especially as your business grows.

AI can help you maintain this tracking informally. After each client meeting or significant conversation, dictate a quick summary of what was discussed and what was decided, and ask the AI to format it as a brief contact note. Save these notes in a simple folder organized by client. Over time, this gives you a reference that helps you remember the details that make clients feel valued when you bring them up.

Using AI for Client Onboarding

The start of a project is a critical moment in the client relationship. Clients who understand what to expect, who their point of contact is, how decisions will be made, and what the communication process will be start the project with confidence rather than anxiety.

A brief project kickoff document, one to two pages, that answers these questions sets the tone for the entire project. AI can generate this document from your project notes and your standard procedures in minutes. It is one of the highest-return investments of time in a client relationship.

Managing Multiple Client Personalities

Different clients want different things from their contractor. Some want detailed technical updates. Some want high-level summaries. Some prefer text messages; others expect formal emails. Some call at all hours; others never reach out unless they have a specific question.

AI can adapt its communication style based on what you know about a client. When you prompt for a client update, tell the AI something about the client's communication style: "This client is very detail-oriented and technical. She will appreciate specifics about what was done and what materials were used." Or: "This client is very busy and prefers brevity. He wants to know whether the project is on track and whether he needs to do anything. Nothing more."

The AI adjusts accordingly. Over time, you develop a brief profile of each active client that you include in relevant prompts, and the communications you send are tailored appropriately without requiring you to rewrite them from scratch.

KEY TAKEAWAYS
Weekly proactive updates eliminate most unsolicited client calls and build referral-driving satisfaction.
Difficult news delivered promptly and clearly is always better than news delivered late or discovered by the client.
Follow-up communications that add value close more bids than those that just ask for a decision.
Tailoring communication style to the client improves relationship quality significantly.

CHAPTER 7

Managing Subcontractors with AI

Subcontractor management is one of the most time-consuming and relationship-sensitive parts of running a GC business. You depend on subs to deliver quality work on schedule, to communicate problems early, and to stand behind their work when issues arise. In exchange, they depend on you for payment, fair treatment, and the information they need to do their jobs.

AI improves several specific parts of this relationship: comparing bids, detecting scope gaps, writing RFIs, managing communication, and documenting performance issues when they arise.

Building a Strong Subcontractor Roster

Your subcontractor roster is one of your most valuable business assets. A reliable network of subs who know your standards, trust your organization, and want to work for you gives you a competitive advantage that is difficult to replicate quickly.

AI helps you build this roster systematically. Qualification requests, the documents you send to new subs to gather their license, insurance, references, and financial information, are standard documents that AI can generate quickly and can be updated by project type or trade.

Comparing Sub Bids: Getting Past the Formatting Problem

Sub bids are notoriously difficult to compare directly. Different subs format their bids differently, include or exclude different items, price allowances differently, and use different terminology for the same work. The effort required to normalize these bids for comparison often means that GCs do a less thorough comparison than they should, especially under time pressure.

AI handles this normalization well. Paste the text of multiple sub bids into your AI tool and ask it to create a side-by-side comparison organized by major scope item, flagging items where one bid differs significantly from the others, and noting items that appear in some bids but not others.

The output gives you a structured comparison that would have taken an hour to prepare manually, in two to three minutes. You can then focus your review time on the differences that matter rather than the formatting work of normalization.

Scope Gap Detection: The Mistake That Costs You

The gap between your project scope and a sub's bid is one of the most expensive problems in construction. The sub bids on what they think is included. You assume their bid covers your full scope. Nobody covers the gap. When the work needs to happen, it becomes a dispute about who is responsible.

AI scope gap analysis is one of the highest-value applications in subcontractor management. Paste your project scope and the sub's bid side by side and ask the AI to identify items in the project scope that do not appear to be covered by the sub's proposal, and items in the sub's proposal that appear to exceed or differ from the project scope.

This analysis is not perfect. AI can miss nuanced scope inclusions or fail to recognize trade-specific terminology. But it catches the obvious gaps consistently, and the obvious gaps are usually the expensive ones.

Writing RFIs That Actually Get Answered

With Requests for Information, the quality varies enormously. A well-written RFI states the issue clearly, references the specific drawing or specification, asks a precise question, and provides the information needed to give a useful answer. A poorly written RFI creates more confusion than it resolves.

AI is good at structuring RFIs. Give it the description of what is unclear, the relevant drawing references, and the specific question you need answered, and ask it to format a professional RFI document. The resulting document is clearer and more complete than most hand-written RFIs, which means you get more useful responses faster.

Sub Communication Templates That Save Hours

A significant portion of sub communication is repetitive: reminder emails about schedule, requests for updated insurance certificates, notifications about site access changes, instructions about daily coordination. AI makes it fast to produce professional versions of all of these from simple prompts.

The investment that pays the most here is building a small library of your most common sub communication templates. Spend an hour with your AI tool generating professional versions of your ten most common sub communications. Save the prompts. Then, when you need to send one of these communications, you are editing a template rather than writing from scratch.

Tracking Lien Waivers and Payment Compliance

Lien waiver tracking is one of those administrative tasks that is critically important but easy to let slide when you are busy. AI can help you maintain a tracking system and draft the waiver requests that you need to send at each payment milestone.

The practical approach: at each payment cycle, paste your sub list into the AI and ask it to generate individual lien waiver request emails for each sub, referencing the specific project, the payment amount, and the waiver type required. This takes five minutes instead of thirty.

Managing Sub Performance

When a sub is underperforming, the instinct is often to avoid the conversation until the situation becomes undeniable. This is understandable but expensive. Problems that could have been addressed in a professional letter often become disputes, replacements, or delay claims.

AI helps you write performance documentation that is factual, professional, and legally defensible. A performance concern letter that documents specific observations, their impact on the project, and what you need the sub to do differently is very different from an angry email written in the heat of the moment. The former is a management tool. The latter can become a liability.

Building Long-Term Sub Relationships

The best subs have choices about who they work for. They gravitate toward GCs who pay on time, communicate clearly, have organized projects, and treat their people with respect. AI helps you be that GC more consistently, because the administrative and communication work that makes you organized and professional becomes sustainable even when you are busy.

Pay attention to the little things that subs notice: prompt responses to their questions, clear scope documents that prevent disputes about what is included, accurate schedules that give them useful lead time, and documentation of their good work that they can use for future references. AI can help with all of these.

PROMPT TO TRY
Compare the following subcontractor bids for [scope of work]. Create a side-by-side comparison organized by major scope items. Flag any items where one bid is more than 15% different from the others. Note any scope items included in some bids but not others. Summarize the key differences I should discuss with each sub before making a selection.
[Paste bid excerpts from each sub]

CHAPTER 8

Site Documentation, Photos, and Compliance

Documentation is insurance. Not metaphorical insurance, though it functions that way, but actual protection against the disputes, claims, and misunderstandings that cost contractors money every year. The GC with thorough, consistent documentation wins arguments. The GC without it relies on memory and trust, which are poor defenses when things go wrong.

The challenge with documentation is time. Good documentation takes fifteen to thirty minutes per day. Most contractors have that time in theory and do not have it in practice, because the end of a workday involves a phone full of unread messages, a subcontractor who needs a decision, and the drive home. AI reduces the friction enough that consistent documentation becomes realistic rather than aspirational.

The Five-Minute Daily Log

The five-minute daily log is the single most valuable documentation habit you can build. Every day you are on site, spend five minutes at the end of your day dictating or typing your observations: who was on site and in what numbers, what work was performed, what the weather was, whether there were any issues or changes, and any decisions that were made.

Send this to your AI tool with the prompt: "Convert these field notes into a professional daily construction log organized by:

weather and site conditions, personnel and equipment on site, work performed by trade and location, issues or delays, decisions or instructions given, changed conditions, and work planned for tomorrow." Save the output.

The result is a professional log that documents the day comprehensively, in a format that holds up in disputes and supports payment applications. The time investment is five minutes of dictation plus thirty seconds of sending the prompt. The payoff, if you ever have a dispute, can be measured in tens of thousands of dollars.

Photo Documentation: The Difference Between Proving and Arguing

Photos are your most powerful documentation tool in construction. A photo of existing conditions before you start. A photo of blocked-out rough work before it is closed in. A photo of the condition that necessitated the change order, taken on the day it was discovered. These photos turn "he said, she said" into evidence.

The problem with photo documentation is not that contractors do not take photos. It is that they take photos without context, in disorganized batches, without descriptions that make them searchable and useful later. When a dispute arises six months after a project closes, a folder of 200 unnamed photos from the project is nearly useless.

AI helps with this in two ways. First, tools like CompanyCam automatically organize photos by project and location, and some AI-assisted tools can generate captions from the photo content. Second, you can dictate brief descriptions of your photos at the end of each day, describing what each set of photos shows and why it was taken, and ask the AI to format this into a photo log that links descriptions to your photo system.

The disciplines that make photo documentation valuable:

- Date-stamped photos are automatic on smartphones. Make sure the timestamp is accurate.
- Before and after photos of any existing condition you are working around or in.
- Photos of concealed conditions before they are closed in, especially anything unusual.
- Photos that document the condition that necessitated any change order, taken on the day of discovery.
- Progress photos at regular intervals, tied to your payment application milestones.

Progress Reports That Clients Actually Read

Progress reports exist on a spectrum from documents that are actually read and valued to documents that are filed unread. The ones that get read have three characteristics: they are brief, they are specific, and they tell the client something they want to know.

AI helps you produce this kind of report quickly. Feed it your weekly status notes, your photo log summary, and any decisions pending from the client, and ask it to produce a one-page progress report. The output is a professional document that keeps your client informed and demonstrates your organizational capability.

Daily Logs That Protect You in Disputes

The daily log is the closest thing construction has to a contemporaneous legal record. Courts and arbitrators give significant weight to daily logs because they were created at the time events occurred, before anyone knew there would be a dispute.

A log that documents that you raised a site condition with the client on a specific date, that a decision was made on another date, and that work proceeded according to that decision can resolve a dispute that might otherwise require arbitration. A memory of those events, unsupported by documentation, rarely does.

The AI-assisted daily log workflow described at the start of this chapter produces exactly this kind of documentation. The key is consistency. A log that exists for 40 of the 60 days of a project is less useful than a log that exists for all 60.

OSHA Documentation and Safety Records

OSHA documentation requirements apply to most construction projects and can be a source of significant liability if not properly maintained. AI can help you draft the documents you need, but it cannot verify that your documentation meets current OSHA standards in your jurisdiction. Always verify regulatory requirements directly.

Where AI helps most in safety documentation:

- Toolbox talk content generation, covered in detail in Chapter 9
- Job hazard analysis drafts for specific tasks
- Incident report formatting from field notes
- Safety plan outlines for site-specific safety programs

As-Built Documentation

As-built documentation is often one of the last things completed on a project and one of the first things cut when schedules are tight. It is also one of the most valuable deliverables you can provide, both for the client and for your own records if warranty issues arise.

AI can help you produce professional as-built documentation packages by organizing your field records, generating cover letters, and creating indexes for your documentation sets. The technical as-built drawings themselves require your field measurements and a CAD tool, but the documentation package around them can be largely AI-assisted.

Using OpenSpace for Large Project Documentation

For larger projects where comprehensive progress documentation is required, OpenSpace represents a significant advancement in documentation capability. A field team member walks the site with a 360-degree camera attached to a hard hat, and OpenSpace's AI automatically maps the footage to your project plans, creates a visual timeline of site progress, and allows comparison between any two dates at any location.

This kind of documentation is invaluable on complex projects where progress disputes, delay claims, or quality issues might arise. The site walk takes fifteen minutes. The resulting documentation package is comprehensive in a way that would take hours to create manually and would still be less complete.

The cost of OpenSpace starts at a few hundred dollars per month and scales with project volume. For GCs managing projects above $2 million or projects with significant delay or dispute risk, the documentation value typically justifies the cost on a single project.

KEY TAKEAWAYS
Daily logs created at the time events occur are the most defensible documentation in disputes.
Photo documentation is only useful if photos have context, description, and organization.
Consistent documentation throughout a project is significantly more valuable than sporadic documentation.
AI reduces the friction of documentation enough to make consistent practice realistic.

CHAPTER 9

Risk, Safety, and Code Compliance

Safety and compliance are the areas where AI provides real practical value while also requiring the most careful application of professional judgment. The tools in this chapter can save you hours of preparation time and improve the quality of your safety documentation. They cannot replace the professional expertise, site-specific knowledge, and regulatory awareness that makes a safety program actually effective.

Read this chapter with that context in mind. The goal is to use AI to make your existing safety practices better and more consistent, not to replace the knowledge that safety depends on.

The Caveat That Comes First

AI safety content must always be verified against current regulatory requirements and adapted for your specific site conditions. OSHA requirements change. Local jurisdictions add requirements. Project-specific conditions create hazards that a general AI response will not account for.

The practical rule: AI drafts safety content, you verify it against current standards and adapt it for your specific situation before it is used with your crew. This review step is not optional. Toolbox talks and hazard analyses that contain inaccurate regulatory information or miss site-specific hazards are worse than no documentation at all.

With that said, here is where AI adds genuine value in safety and compliance.

Pre-Task Hazard Identification

Before any non-routine task begins, a pre-task hazard analysis identifies the potential hazards, the controls to address them, and any specific PPE or permits required. For tasks you do repeatedly, these analyses can be templated. For unusual tasks, they need to be specific to the work.

AI can generate a comprehensive first draft of a hazard analysis for any described task in about a minute. The draft identifies common hazards for that type of work, suggests standard controls, and flags any tasks that require permits or special procedures. Your superintendent or safety lead then reviews the draft, adds site-specific conditions, and uses it for the pre-task briefing.

The value is not that AI replaces the safety conversation. It is that it ensures the conversation starts from a comprehensive foundation rather than from memory. Hazards that experienced workers take for granted are sometimes the most dangerous, because they stop getting mentioned.

Site-Specific Safety Plans

Most GCs need a site-specific safety plan for every project above a certain threshold of size or complexity. These plans typically cover site access and egress, emergency response, hazard identification, personal protective equipment requirements, and specific procedures for high-hazard activities present on the project.

Generating these plans from scratch for every project is time-consuming. AI can produce a comprehensive outline for a site-specific safety plan based on your project description in minutes. You then fill in the site-specific details, verify the regulatory requirements, and review the finished plan with your site team.

The result is a more thorough safety plan produced faster than starting from a blank document or a generic template.

Toolbox Talk Content: Weekly Safety Meetings Made Easy

Toolbox talks are one of the highest-value safety practices available to GCs. A consistent weekly safety meeting keeps hazards top of mind, demonstrates safety culture to your team, and creates a documentation record that supports your OSHA compliance posture.

The challenge is content. Coming up with a relevant, engaging ten-minute toolbox talk every week, for months or years, is a creative task that most field supervisors approach with dread. AI solves this completely.

Give the AI a safety topic relevant to your current work, the format you want, and any specific site conditions to address, and it will produce a complete toolbox talk including the main points, the OSHA references, and the review question in about thirty seconds. The resulting content is more comprehensive and better organized than most manually written talks.

PROMPT TO TRY

Generate a toolbox talk on [safety topic] for a construction crew. Format it as a ten-minute presentation. Include: why this topic matters (keep it real and practical), main hazards workers face, how to recognize warning signs, specific protective measures and controls, any applicable OSHA requirements in plain language (note: verify against current standards), and a brief knowledge check question. Use plain, direct language.

Topics to rotate through over a quarter: fall protection, struck-by hazards, electrical safety, heat illness, silica exposure, ladder safety, eye protection, hand protection, caught-in/between hazards, housekeeping, and emergency response procedures. A quarter's worth of toolbox talks in thirty minutes of AI prompting.

Code Compliance: AI as a Research Starting Point

Code research is one of the most time-consuming and high-stakes parts of estimating and project management. Getting it wrong can mean rework, failed inspections, and project delays. Getting it right requires knowing which codes apply, what the current versions require, and how your local jurisdiction interprets them.

AI is genuinely useful for initial code research and for understanding the structure of code requirements, while being genuinely limited when it comes to the specific, current, locally-adopted version of any code. Use AI to get oriented, then verify against the actual adopted code in your jurisdiction.

Good uses of AI for code research: understanding the structure of a code section, getting a general sense of what a type of project typically requires, identifying what questions to ask your inspector or building department, and understanding the intent behind a code requirement when the language is unclear.

Not good uses: relying on AI for specific dimensional requirements, fire-rating requirements, energy code compliance details, or any other specific code value that you will be held to in an inspection.

Insurance Documentation and Claims Support

When an incident occurs on site that may result in an insurance claim, the quality of your documentation in the hours and days immediately following the event significantly affects the outcome. AI can help you produce professional incident reports, chronologies, and correspondence in a format that supports your claim rather than undermining it.

The important caveat on insurance documentation is the same as everywhere in this book: AI produces drafts that you review. Every fact in insurance-related communications needs to be accurate. Do not let AI's confident tone lead you to submit a report that contains inaccuracies.

Safety Culture Beyond the Paperwork

The most important safety outcome, the actual prevention of injuries, comes from safety culture, not from paperwork. AI helps you produce the documentation that supports a safety program, but the culture comes from leadership behavior on the site.

The contractors who have the best safety records are the ones where every person on the job, from the principal to the newest laborer, knows that unsafe behavior will be addressed immediately and that raising a safety concern is expected and respected. No AI tool creates that culture. Leadership does.

What AI does is remove the administrative burden that sometimes gets in the way of safety focus. When generating a toolbox talk takes thirty seconds instead of two hours, you do them every week. When hazard analyses are easy to produce, you do them for every non-routine task. The documentation habit supports and reinforces the culture.

KEY TAKEAWAYS
AI generates safety content drafts that must be verified against current regulatory standards.
Weekly toolbox talks are one of the highest-value safety practices - AI eliminates the content preparation challenge.
AI is useful for initial code research orientation but not for specific dimensional or compliance values.
Safety culture comes from leadership behavior. Documentation supports culture but does not create it.

CHAPTER 10

Your AI Prompt Library

This chapter is your reference section. Every prompt here has been written specifically for construction scenarios and can be used directly in ChatGPT, Claude, or Gemini without modification, except for filling in the details in brackets. Copy the ones you use most often to a note on your phone so they are available in the field.

The prompts are organized by workflow so you can find what you need quickly. When you use a prompt for the first time, the output will give you a clear sense of how much you need to customize it for your specific situation. Over time, you will develop your own variations that work best for your company and your project types.

Estimating and Scope Analysis

Scope Gap Analysis

PROMPT TO TRY
Review the following project scope and give me three things: (1) a list of items that are ambiguous or could reasonably be interpreted differently by different contractors, (2) a list of items commonly included on this type of project that are not explicitly addressed in this scope, and (3) items I should explicitly exclude from my bid to prevent scope creep. Format each as a separate numbered list.
[Paste scope of work here]

Budget Line Items by Trade

PROMPT TO TRY

I am estimating a [project type and size] in [location]. Generate a comprehensive list of budget line items organized by trade division. For each trade, list the primary scope items a GC should account for. Flag any items that are commonly underestimated or overlooked on this type of project with an asterisk.

Bid/No-Bid Analysis

PROMPT TO TRY

Help me evaluate this bid opportunity. Here are the details: project type [describe], client type [owner, developer, tenant, public agency], estimated value [range], bid due date [date], location [city/state], who else is bidding [known competitors], our current backlog [light/moderate/heavy]. Give me three strong reasons to pursue this bid and three reasons to pass, then make an overall recommendation.

Historical Cost Calibration

PROMPT TO TRY

I have historical cost data from [number] completed projects of this type. The average actual costs per square foot by trade were: [list your data]. For the new project I am estimating, I am seeing these preliminary estimates per square foot: [list]. Identify any trades where my current estimate is significantly different from my historical data and suggest whether the difference is likely justified or worth investigating further.

Risk Flag Review

PROMPT TO TRY

Review the following bid package and identify the top five risk factors I should price into my bid or flag as exclusions. Consider: site conditions, contract terms, scope complexity, client type, timeline, and any unusual specification requirements.

[Paste key sections from bid package]

Proposals and Contracts

Complete Project Proposal

PROMPT TO TRY

Write a professional project proposal for a [client type] hiring us for [project description]. Organize it with these sections: Project Understanding (demonstrate we grasp their goals), Our Approach (how we will execute), Scope of Work (specific inclusions), Exclusions (what is not included), Timeline (key milestones and completion date), Our Team (brief relevant credentials), and Investment (how they authorize the work). Two pages maximum. The tone should be professional but approachable.

[Add your bullet points about the project here]

Contract Risk Review

PROMPT TO TRY

Review the following contract clauses for a construction project. Flag anything that is: (1) outside standard industry practice, (2) unusually one-sided in favor of the owner, or (3) likely to create disputes during project execution. For each flagged item, explain what the concern is and what alternative language I might propose. Focus on: payment timing, retainage, change order procedures, indemnification, consequential damages, termination rights, and lien provisions.

[Paste contract sections]

Change Order Narrative

PROMPT TO TRY

Write a professional change order narrative. Original scope did not include: [describe]. Reason this work is now required: [explain]. Additional work involved: [describe]. Cost impact: [amount]. Schedule impact: [days]. Write this as a factual business communication that makes the case for approval clearly and professionally. Do not be defensive. Reference the original scope language where relevant.

Scope of Work Document

PROMPT TO TRY

Write a detailed scope of work document for a [project type]. Include: (1) Work Included section with specific items by trade, (2) Allowances section listing items priced at a preliminary allowance with specific allowance amounts, (3) Exclusions section with explicitly excluded items, and (4) Owner-Provided Items section if applicable. Make the language specific enough to prevent scope disputes.

[Add your notes about what is and is not included]

Letter of Intent Response

PROMPT TO TRY
Write a professional response to a letter of intent from an owner who wants to begin preliminary work before contracts are signed. Acknowledge the LOI, confirm the scope of preliminary work we will proceed with and the dollar limit, state that all work beyond this limit requires an executed contract, and include appropriate reservation of rights language. Tone should be positive and cooperative while protecting our interests.

Scheduling and Timeline Management

Project Schedule Framework

PROMPT TO TRY
Generate a project schedule for the following project. Include logical sequencing with dependencies noted, estimated durations for each phase, key milestones, long-lead items that require early ordering or coordination, and any activities where sequencing problems commonly occur.
Project type: [describe]
Trades involved: [list]
Start date: [date]
Required completion: [date]
Known constraints: [phasing, occupancy, inspections, owner milestones]

Two-Week Look-Ahead

PROMPT TO TRY
Create a two-week look-ahead schedule. Format it by day for the next two weeks. For each day, list the trades and activities planned, any material deliveries or inspections expected, and any owner or design team decisions needed.
Current project status: [describe where you are]
Active trades: [list]
Work completed last week: [summary]
Planned work next two weeks: [bullet points]
Deliveries expected: [list]
Inspections scheduled: [list]

Delay Notice to Owner

PROMPT TO TRY
Write a formal delay notice to an owner for a construction project. Include: the specific cause of the delay and when it was identified, the anticipated impact on the project completion date in days, the steps we are taking to minimize the impact, any decisions or actions required from the owner, and a statement reserving our rights under the contract for any delay-related costs. The tone should be professional and factual.

Recovery Schedule Analysis

PROMPT TO TRY
Our project is currently [X] days behind the contract schedule. The original completion date was [date] and the current projected completion is [date]. We have the following resources available for recovery: [list subs, crew availability, budget for overtime]. Generate five potential recovery strategies ranging from minimal additional cost to aggressive recovery. For each strategy, describe the approach, the realistic days of recovery achievable, and the main risk or tradeoff.

Client Communication

Weekly Client Update

PROMPT TO TRY
Write a weekly project update email for a client. Three short paragraphs maximum. Professional but warm tone.
Project: [brief description]
Completed this week: [bullet points]
Planned next week: [bullet points]
Decisions needed from client: [list or "none at this time"]
Items affecting schedule or budget: [list or "none"]
Milestone or good news to highlight: [optional]

Difficult News Communication

PROMPT TO TRY
Write a professional email to a client explaining a difficult situation. Organize it using this structure: (1) a clear statement of the situation without minimizing or dramatizing, (2) the cause to the extent we know it, (3) specifically what we are doing to address it, and (4) what the client needs to know or do and what happens next.
Situation: [describe clearly]
Cause: [explain]
Our response: [list actions taken or planned]
Client action needed: [describe or state "none required"]

Proposal Follow-Up Email

PROMPT TO TRY
Write a follow-up email to a client who received our proposal [X] days ago for a [project type]. Reference something specific about their project or goals to show this is not a generic follow-up. Add one piece of value (a related thought, a clarification, or a relevant example). Gently invite them to ask questions or schedule a call. Keep it to three short paragraphs. Not pushy.

Project Closeout Letter

PROMPT TO TRY
Write a professional project closeout letter for a client. Include: acknowledgement that the project is complete, summary of the warranty coverage and how to make a warranty claim, who to contact for service, a brief statement of appreciation for their business, and an invitation to discuss future projects or provide a referral if they are satisfied. Warm and professional tone.

Subcontractor Management

Sub Bid Comparison

PROMPT TO TRY
Compare the following subcontractor bids for [scope of work]. Create a side-by-side comparison organized by major scope items. Flag any items where one bid is more than 15% different from the others. Note any scope items included in some bids but not others. Summarize the key differences I should discuss with each sub before making a selection.
[Paste bid excerpts from each sub]

Scope Gap Detection

PROMPT TO TRY
Compare my project scope to this subcontractor's proposal. Identify: (1) items in the project scope that do not appear to be covered by the sub's proposal, and (2) any items in the sub's proposal that appear to exceed or differ from the project scope. Format as two separate lists.
Project scope: [paste]
Sub proposal: [paste]

RFI Draft

PROMPT TO TRY
Draft a Request for Information for a construction project. Format it professionally with fields for project name, RFI number, date, requested response date, subject, description of the issue, specific question, reference to relevant drawing or specification section, and space for response.
Issue description: [describe what is unclear or conflicting]
Plan/spec reference: [sheet number or specification section]
Specific question: [what you need to know]
Response needed by: [date]

Performance Concern Letter

PROMPT TO TRY
Write a professional letter to a subcontractor documenting a performance concern. Include: specific description of the issue (what was observed and when), the impact on the project, what we need from the sub to address the issue and by what date, and a statement that this is being documented and that continued performance issues may affect their standing on this and future projects. Factual and professional, not threatening.

Sub Qualification Request

PROMPT TO TRY

Write a subcontractor qualification request letter for a [trade] subcontractor we are considering for [project type] projects. Ask for: current license and insurance certificates, three references from projects of similar type and value completed in the last three years, their current bonding capacity and bonding company contact, their safety record including EMR for the past three years, their financial stability including a brief statement of their current work capacity, and their key project team members who would be assigned to our work.

Safety and Documentation

Toolbox Talk

PROMPT TO TRY

Generate a toolbox talk on [safety topic] for a construction crew. Format it as a ten-minute presentation. Include: why this topic matters (keep it real and practical), main hazards workers face, how to recognize warning signs, specific protective measures and controls, any applicable OSHA requirements in plain language (note: verify against current standards), and a brief knowledge check question. Use plain, direct language.

Job Hazard Analysis

PROMPT TO TRY
Generate a Job Hazard Analysis for the following task. Format it as a three-column table: Task Step, Potential Hazards, Recommended Controls.
Task: [describe the specific work]
Work environment: [indoor/outdoor, heights, confined space, etc.]
Workers: [number and experience level]
Equipment: [tools and equipment involved]
Known hazards: [any you are already aware of]

Daily Construction Log

PROMPT TO TRY
Convert the following field notes into a professional daily construction log. Include clearly labeled sections for: date and weather conditions, trades on site with crew counts, work accomplished by trade and location, issues or delays encountered, decisions made or instructions given, changed conditions or unexpected discoveries, visitor log if applicable, and work planned for the following day.
[Paste voice transcription or field notes]

Incident Report

PROMPT TO TRY

Draft an incident report for the following event. Format it professionally with sections for: incident date, time, and exact location; description of what happened in sequence; names and roles of all parties involved; nature of any injury or property damage; immediate first aid or emergency response provided; witnesses; and corrective actions implemented or planned. Factual, specific, and complete.

[Describe the incident with all known facts]

Site-Specific Safety Plan Outline

PROMPT TO TRY

Generate an outline for a site-specific safety plan for a [project type] project. Include all major sections a complete safety plan should contain, with a brief description of what each section should cover. I will use this as a framework to complete the full plan.

Business and Operations

Client Reference Request

PROMPT TO TRY

Write a professional email asking a satisfied client to serve as a reference for our company. Include: a brief statement of appreciation for working with them, a specific mention of something that went well on their project, what serving as a reference entails (a phone call from a prospective client or a written testimonial), and how much we would appreciate their support. Warm and genuine, not transactional.

Team Meeting Agenda

PROMPT TO TRY

Create a structured agenda for a [weekly/monthly] project team meeting. Include standard sections that a GC's team should cover in every meeting: safety moment, active project status by project, upcoming bids and proposals, staffing and resource planning, client issues or escalations, sub performance items, and any other business. Estimated time allocations for each section.

New Employee Orientation Checklist

PROMPT TO TRY

Generate a comprehensive new employee orientation checklist for a construction company. Include items for: company policies review, safety orientation, site-specific hazard training, required PPE, communication procedures, emergency response, payroll and HR information, and any trade-specific orientation items. Format as a printable checklist with signature lines for employee and supervisor.

Vendor Evaluation

PROMPT TO TRY

Help me evaluate a new material supplier or vendor I am considering adding to our approved vendor list. Generate a list of ten questions I should ask them to assess: their financial stability, their ability to meet our typical order volumes and lead times, their quality control processes, their return and warranty policies, their communication and service standards, and their pricing structure compared to alternatives.

CHAPTER 11

The Best AI Tools for General Contractors

The AI tool landscape moves fast, and by the time you read this, new tools will have appeared and some existing ones will have changed their pricing or features. The goal of this chapter is not to provide a definitive ranking that will be outdated in six months, but to give you a clear framework for evaluating tools and an honest assessment of the leading options available as of the time this book was written.

One more note before the tools. Verify current pricing and features directly with each vendor before you spend money. Subscription pricing in software changes frequently, and what is listed here may not match what you find when you go to sign up. Treat the pricing information here as an order-of-magnitude guide, not a specific number to hold the vendor to.

Framework for Choosing AI Tools

Before you evaluate any specific tool, answer these four questions about your business. They will dramatically narrow the field and help you avoid spending time evaluating tools that are not a fit.

- Where is your biggest pain point? The tool that solves your most painful daily problem will always deliver more ROI than the theoretically superior tool that solves a problem you do not have as often.

- What is your realistic budget? AI tools range from free to several hundred dollars per month per user. Small GCs should start with free-tier tools and only upgrade when they have confirmed the value.
- What tools do you already use? The best AI tool for your business might be AI features added to software you already pay for. If you are on Procore, explore its AI features before adding a separate tool that requires data integration.
- How much time can you realistically invest in learning? Some tools have meaningful learning curves. Others are designed to be picked up in an afternoon. Be honest about how much time you and your team actually have for tool adoption.

General-Purpose AI Tools

These are the foundation of any contractor's AI toolkit. You will use one of these every day for drafting, reviewing, analyzing, and communicating.

Tool	What It Does	Cost	Best Fit
ChatGPT (OpenAI)	General drafting, analysis, document review, Q&A. The most widely used AI platform globally with the largest ecosystem of add-ons and integrations.	Free / Plus $20/mo / Team $25/user/mo	All contractors starting with AI
Claude (Anthropic)	Especially strong for long document handling and clear writing. Preferred by many contractors for contract review and	Free / Pro $20/mo / Team $25/user/mo	Document-heavy workflows

Tool	What It Does	Cost	Best Fit
	detailed proposal writing.		
Gemini (Google)	Integrates directly with Gmail, Docs, and Drive. Best choice for contractors already operating in Google Workspace.	Free / Advanced $19.99/mo included with Google One AI Premium	Google Workspace users
Microsoft Copilot	Integrates with Microsoft 365 tools including Word, Excel, Outlook, and Teams. Best for contractors using the Microsoft ecosystem.	Free limited / M365 Copilot $30/user/mo	Microsoft 365 users

AI Estimating and Takeoff Tools

These tools address the most time-consuming part of the bidding process: getting quantities from plans and assembling preliminary costs.

Tool	What It Does	Cost	Best Fit
Togal.AI	AI-powered plan takeoff that automatically extracts quantities from PDF drawings. Claims up to 98% accuracy and 5x faster takeoff. Togal.CHAT lets you talk to your plans with natural language questions.	Paid subscription (contact for pricing)	Commercial GCs, larger residential builders
STACK	Cloud-based estimating with AI quantity suggestions and access to cost databases. Free version handles basic takeoff; paid tiers add integrations and collaboration features.	Free / Pro and Enterprise tiers	GCs of all sizes, especially those new to digital estimating

Tool	What It Does	Cost	Best Fit
Beam AI	Designed specifically for small to mid-size contractors. AI drafts estimates from scope descriptions and site notes, and learns from your cost history over time.	Paid subscription	Small to mid-size GCs
Handoff	Residential-focused AI estimating. Trained on 100,000+ real residential projects with ZIP-code-level pricing from Home Depot, Lowe's, and other suppliers.	Subscription with free trial	Residential contractors and remodelers
PlanSwift (Trimble)	Takeoff software with AI-assisted features and strong Excel integration. Established tool with a large user base.	Paid license or subscription	GCs and specialty contractors with existing Excel workflows
ProEst (Autodesk)	Cloud-based estimating with AI suggestions, cost databases, and Autodesk ecosystem integration.	Paid subscription	Mid to large GCs already using Autodesk products

AI-Enhanced Project Management

These platforms bring AI capabilities to the broader project management workflow, including scheduling, RFIs, submittals, budget tracking, and field coordination.

Tool	What It Does	Cost	Best Fit
Procore with AI	The dominant construction management platform now includes AI capabilities for RFI automation, schedule analysis, budget variance alerts, and document processing.	Paid (enterprise pricing by project volume)	Mid to large GCs already on or considering Procore
Autodesk Construction Cloud	Integrates AI with BIM for clash detection, design coordination, and automated risk analysis. Best for projects where design coordination across disciplines is complex.	Paid (enterprise pricing)	GCs on complex projects with active design coordination
SmartPM	AI-powered schedule analytics and project controls. Specializes in schedule diagnostics, delay analysis, and claims mitigation support.	Paid subscription	GCs managing schedules on complex projects or with delay risk
Buildots	360-degree camera system with AI that compares site progress to BIM plans automatically. Reduces manual reporting by automating progress tracking.	Paid (hardware + software)	GCs on projects where progress tracking and documentati on are intensive

Field Documentation Tools

Tool	What It Does	Cost	Best Fit
CompanyCam	Photo documentation with automatic organization by project and location, AI captioning, and report generation. The most widely adopted field photo tool in construction.	Free limited / Unlimited from $19/user/mo	All GCs with active field teams
OpenSpace	360-degree site capture with AI progress tracking against plans. More comprehensive documentation than photo-by-photo methods.	Paid subscription (contact for pricing)	GCs on larger projects requiring comprehensive progress documentation
Fieldwire	Site management and task tracking with photo documentation. Simpler than Procore but more powerful than basic note-taking.	Free limited / Pro and Business tiers	Smaller GCs wanting project management with field documentation
PlanRadar	Mobile-first documentation and inspection platform with AI features for defect tracking and compliance reporting.	Paid subscription	GCs managing multiple projects with QA/QC requirements

Specialized Contract and Document AI

Tool	What It Does	Cost	Best Fit
Document Crunch	AI specifically trained on construction contracts. Analyzes contracts for risk, flags non-standard clauses, and provides plain-language explanations of complex provisions.	Paid subscription	GCs reviewing significant commercial contracts regularly
Kira (Litera)	Enterprise contract review AI. More powerful than general AI tools for complex contract portfolios, but priced for larger organizations.	Enterprise pricing	Larger GCs with high-volume contract review needs
LawGeex	AI contract review with construction-specific modules. Similar positioning to Document Crunch.	Paid subscription	GCs wanting purpose-built contract review AI

Building Your Tool Stack: A Practical Approach

A common mistake when starting with AI tools is trying to implement too many things at once. The better approach is to build incrementally, confirming value at each step before adding the next layer.

Starting Stack: Month 1 - One general-purpose AI tool, free tier. Use it daily for client communication, email drafting, and scope review. Cost: free. Time investment: ten minutes per day to build the habit.

Growth Stack: Months 2 to 4 - Add one specialized estimating tool on a trial basis. Apply it to your next two or three bids and track the time savings. Add CompanyCam if you do not already have a photo documentation system. Cost: $50 to $200 per month depending on selections.

Mature Stack: Month 6 and Beyond - Upgrade general AI to paid tier for higher usage limits. Keep or replace specialized estimating tools based on trial results. Consider adding project management AI features if your current PM platform offers them. Cost: $200 to $500 per month depending on company size and tool selections.

THE MOST IMPORTANT TOOL SELECTION RULE

The best AI tool stack is the one you actually use consistently. A free tool you use every day will always outperform an expensive tool that sits unused because the learning curve was too steep or the workflow did not fit how you actually work. Start with the problem that hurts the most. Solve that one first.

CHAPTER 12

Ethics, Liability, and Data Privacy

This is the chapter that most AI books skip, or mention briefly in a disclaimer and move on from. We are treating it as a full chapter because it covers things that every contractor needs to understand before deploying AI tools on real projects with real clients, real money, and real professional liability attached.

None of this chapter is legal advice. It is practical guidance based on how AI tools actually work and the specific ways they can create problems if used without understanding their limitations.

You Are Still the Professional

The most important thing to understand about AI in construction is that it does not change who is responsible for the work. The contractor is responsible for the estimate, the proposal, the scope document, the change order, and the project outcome. The fact that AI helped produce a document does not transfer any portion of that responsibility to the AI tool or its maker.

This is not just a legal point. It is a practical one. AI tools produce output that sounds confident and professional but is not always accurate. An AI-generated scope document that misses a critical exclusion is your scope document, with your liability attached. An AI-drafted change order narrative that contains a factual error is your change order narrative.

The practical rule that runs through this entire book is the right answer to the liability question: AI produces drafts that you review. You bring your professional knowledge, your site-specific context, and your liability. AI brings speed and consistency. The combination works. Relying on AI without review does not.

Specific Liability Considerations for Contractors Using AI

Estimating: AI estimating tools produce quantities and costs that require professional review. A takeoff quantity that is wrong because the AI misread a plan is your error, not the tool's. Before relying on any AI-generated quantity for a bid, verify the key quantities with your own review.

Contract language: AI can help you draft contract clauses and scope language, but contracts have legal consequences. Non-standard language that AI generates may not mean what you think it means or may create unintended obligations. Have your attorney review any new contract language before using it.

Code compliance: AI can summarize code requirements and help you research applicable standards. It cannot verify that what it says matches the current, locally-adopted version of any code. Never use AI as your sole source for code compliance determinations.

Safety documents: AI safety content must be verified against current OSHA standards and adapted for your specific site conditions. A toolbox talk that references an outdated OSHA standard or misses a site-specific hazard creates a documentation problem, not a documentation asset.

Data Privacy: What You Share When You Use AI

When you paste information into a general-purpose AI tool like ChatGPT or Claude, that information is processed by the tool's servers. Understanding what happens to that information is important for protecting your clients and your business.

The basic principle: Treat AI tools like any other cloud service. Information you put in is stored on someone else's servers. For most construction workflows, this is not a meaningful risk. Drafting a proposal, reviewing a scope document, or generating a toolbox talk does not involve information that requires extraordinary protection.

Where it matters more: Client financial information, proprietary bidding strategies, personnel information, and sensitive legal or insurance matters deserve more careful handling. For these situations, consider whether the AI tool's data handling practices meet your risk tolerance.

Practical guidance:

- Use the paid, commercial versions of AI tools rather than free consumer versions. Commercial versions typically have stronger data protection terms.
- Review the data privacy policies of the tools you use, especially for specialized tools that handle your project data.
- Do not paste client Social Security numbers, financial account information, or other sensitive personal data into general-purpose AI tools.
- For projects with NDAs or confidentiality requirements, review whether your AI tool usage is consistent with your contractual obligations.

When to Override AI

There are situations where you should not use AI output at all, regardless of how good it looks.

When the facts are wrong: If you know the AI has generated content with factual errors and you cannot fix all of them, start over rather than try to patch a flawed draft.

When the stakes require expert review: Legal documents being executed, insurance claim statements, safety incident reports for serious injuries, and communications about potential litigation should be reviewed by the appropriate professional, not just by you after AI drafting.

When the AI misunderstands your situation: If a prompt response shows that the AI does not understand your specific situation well enough to be useful, reprompt with more context or address the communication manually.

The Insurance Conversation You Should Have

Talk to your insurance broker about how your current coverage applies to AI-related issues. The questions worth asking:

- Does my professional liability coverage apply if an AI-generated document I produced contains an error that leads to a claim?
- Do my data privacy protections apply to client information I handle through AI tools?
- Are there specific AI tool uses that my current coverage does not address?

This conversation takes thirty minutes and may identify gaps in your coverage before they become problems.

The Ethics of Using AI in Client Work

The practical ethics of AI in construction are largely extensions of the professional ethics that good contractors already practice.

Disclose when relevant. If a client asks whether AI was used in producing a document, answer honestly. If your use of AI is a significant part of your efficiency advantage, there is no reason to obscure that.

Do not pass off AI output as human expertise. AI can draft a proposal, but it does not know your team, your track record, or your approach to client relationships. The proposal should represent your actual capabilities, not an AI description of a generic contractor.

Review everything before it goes to a client. Not just for errors, but for appropriateness. AI does not know the nuances of your client relationships. A technically correct communication can still be wrong in tone for a specific client or situation.

ETHICS AND LIABILITY CHECKLIST
1. Review everything AI produces before using it on a real project.
2. Verify code compliance requirements directly with your local authority having jurisdiction.
3. Be thoughtful about what client information you put into AI tools.
4. Have important contracts reviewed by an attorney.
5. Talk to your insurance broker about AI implications.
6. Take professional responsibility for everything you sign.

CHAPTER 13

Your 90-Day AI Implementation Plan

The biggest mistake contractors make when adopting new tools is trying to change everything at once. They install five apps in a weekend, spend a week trying to figure out how they all work together, and then go back to the way they were doing things before because the disruption was more painful than the problem they were trying to solve.

This chapter lays out a deliberate, phased approach to building AI into your business in a way that actually sticks. Ninety days is the right timeframe because it is long enough to build real habits and see measurable results, but short enough to stay motivated and adjust course before you are too deep into a direction that is not working.

Before You Start: Setting Yourself Up for Success

A few things to do before your ninety days begin.

Measure Your Baseline

You cannot know whether AI is helping you if you do not know how long things take today. Before your first day, spend twenty minutes estimating the following:

- How many hours per week do you spend on estimating and bid preparation?

- How many hours per week do you spend on client communication and documentation?
- How many hours per week does your team spend on administrative and documentation tasks?
- How many bids do you submit per month on average, and what is your current win rate?

Write these numbers down. At the end of ninety days, you will compare your reality against this baseline to calculate actual ROI.

Pick Your Starting Tools

Choose one general-purpose AI tool to start with. If you are not sure which one, start with the free version of ChatGPT or Claude. Both are available in your browser and on your phone immediately. Do not subscribe to anything yet. You are building a habit, not a tech stack.

Identify Your Two Starting Workflows

Pick the two workflows that cause you the most pain today. For most contractors, these are estimating and client communication. But your situation might be different. Maybe sub coordination is your biggest headache. Maybe documentation is where you lose the most time. Pick the two that matter most to you.

Days 1 to 30: Build the Habit

The goal of month one is simple: use AI every day for your two chosen workflows, build comfort with how the tool works, and save your best prompts so you can reuse them.

Week 1: Learn the Tool

Do not try to use AI for anything mission-critical in your first week. Use it for low-stakes tasks: draft an email, summarize a document you need to read, generate a list of questions for an upcoming meeting. The goal is to get comfortable with how the tool responds to different types of prompts and what good prompting looks like for your situation.

Spend fifteen minutes each morning for the first week doing something small with your AI tool. By the end of the week, you will have a sense of what the tool is good at and what it struggles with.

Week 2: Apply to Real Work

In week two, start using AI for actual work in your two chosen workflows. Draft your first proposal with AI assistance. Use AI to review a scope document for gaps. Generate a week's worth of client updates with AI help. The quality of the first several outputs will be uneven. That is normal and expected. Keep a note of the prompts that produced the best results.

Weeks 3 and 4: Refine and Standardize

By the end of month one, you should have five to ten prompts that you have tested and refined for your specific situation. Save these somewhere you can access them from your phone. These become the foundation of your personal prompt library.

Also by the end of month one, you should have a clear sense of how much time AI is saving you on your two workflows. Even a rough estimate, such as "I spent about four hours less on estimating this week than I usually do," is enough to tell you whether the investment of time in building this habit is paying off.

Days 31 to 60: Add a Specialized Tool

In month two, add one specialized tool in the area where you are getting the most value from your general AI habit. For most contractors, that will be an estimating tool.

Selecting and Trialing a Specialized Tool

Go back to the tool table in Chapter 11 and identify two or three tools that seem like a fit for your project type and budget. Sign up for a trial on the one that seems best. Most of the tools in the estimating category offer free trials long enough to test them on two or three real bids.

For your trial, apply the tool to your next two live bids. Track the time from plan receipt to bid submission with the tool versus your best recollection of how long this usually takes. Note the accuracy: does the AI-generated estimate require significant correction, or is it reasonably close to what you would have produced manually?

Introducing AI to Your Team

In month two, start sharing your best workflows with one or two people on your team. Show them the specific prompts you have found most useful. Walk them through a simple workflow. Do not try to train everyone on everything at once.

The person to start with is whoever you believe will be most open to trying new tools. Early adopters on your team become internal champions who help bring others along. Early skeptics who are forced to adopt before they are ready become the people who undermine adoption across the rest of the team.

Days 61 to 90: Measure, Decide, and Scale

Month three is about converting your experiment into a deliberate strategy based on real data.

Calculate Your Return on Investment

Go back to your baseline measurements from before day one. Compare them to where you are now. Be as specific as you can:

- Hours saved per week on estimating
- Hours saved per week on communication and documentation
- Number of additional bids submitted per month
- Changes in win rate, if you have enough data to assess
- Cost of the tools you are now paying for

For most contractors who have genuinely built the habit, the math at this point is clearly positive. A project manager saving five hours per week at a fully-loaded rate of $80 per hour has generated $400 per week in recovered time, or about $20,000 per year. A GC who has submitted three more bids per month and won one additional project has generated significantly more than the cost of any tools involved.

Decide What to Keep and What to Drop

Not every tool will earn a permanent place in your workflow. The standard for keeping a specialized tool should be clear and measurable: does it save enough time or reduce enough risk to be worth its monthly cost? If you cannot demonstrate that clearly after two months of use, it is probably not the right tool for your current situation.

Expand to Additional Workflows

If month one and month two have gone well, month three is when you start applying AI to additional workflows. Add field documentation. Add scheduling. Add subcontractor communication. Each additional workflow should follow the same pattern: start with existing AI tools and your prompt library, measure the results, and only add new specialized tools if the existing approach has clear gaps.

Building an AI-Enabled Team Over Time

The goal beyond ninety days is a business where AI-assisted workflows are the normal way of doing things, not a special initiative that depends on one person's enthusiasm. Getting there requires a few additional steps.

Document Your Best Workflows - As you refine what works, write it down in a format that someone else can follow. A one-page description of how your team uses AI for estimating, or a template for weekly client updates, is a company asset that does not disappear when someone changes roles or leaves.

Create a Company Prompt Library - Build a shared document, accessible to everyone on your team, that contains your best prompts organized by workflow. Update it regularly as you refine what works. This is a living document, not a one-time project.

Set Regular Review Milestones - Every six months, take thirty minutes to review your tool stack, your workflow effectiveness, and the tool landscape. New tools emerge regularly. Existing tools add features. Pricing changes. A brief semi-annual review keeps your approach current without requiring constant attention.

Phase	Goal and Actions
Days 1-7: Learn	Get comfortable. Use AI daily for low-stakes tasks. Build basic prompting fluency.
Days 8-30: Apply	Use AI for real work in 2 chosen workflows. Save the prompts that work. Measure time saved.
Days 31-45: Add Tools	Trial one specialized tool on 2-3 real projects. Track time and accuracy.
Days 46-60: Train Team	Share your best workflows with 1-2 team members. Don't train everyone at once.
Days 61-75: Measure ROI	Compare time spent vs. baseline. Calculate actual return. Decide what to keep.
Days 76-90: Expand and Standardize	Apply AI to additional workflows. Document your best processes. Build a company prompt library.

THE 90-DAY SUMMARY
Month 1: Build the habit with two workflows.
Month 2: Add one specialized tool and one team member.
Month 3: Measure what you have built and scale what is working.
Everything starts here.

CHAPTER 14

Real Stories and Real Results

The best evidence that something works is not a statistic or a research finding. It is a story from someone who runs a business like yours, tried something specific, and measured what happened. The following stories are composite profiles drawn from real patterns of AI adoption across small and mid-size GC operations. The specifics have been adjusted and combined, but the outcomes reflect what contractors who are actually using these tools are experiencing.

Story One: The Estimating Bottleneck

Profile: A 6-person GC specializing in commercial tenant improvements, annual revenue approximately $4 million.

Before AI tools, this contractor's biggest constraint was estimating capacity. The principal and one part-time estimator could realistically process two to three bids per week. They were selective about which opportunities they pursued because the estimating time was the bottleneck, not the field capacity.

They started with Togal.AI for plan takeoff and Claude for proposal writing. The first month was an adjustment period. The takeoff tool produced quantities they needed to review and correct, and the proposals needed significant editing to sound like the company's voice. By the end of month two, both tools were producing outputs that required less correction.

By month three, they were processing four to five bids per week consistently. The time savings on takeoff, roughly five to six hours per bid, were being reinvested in reviewing more opportunities. Their bid volume for the quarter was up 60 percent. Their win count for the quarter was up 40 percent, even though their win rate had not significantly changed.

The ROI calculation was straightforward. The two tools combined cost about $300 per month. The incremental revenue from two additional projects per month, at an average project size of $200,000 and a margin of eight percent, produced approximately $32,000 in incremental margin per month. The tool costs paid for themselves in the first week of the first month they were in use.

What surprised this contractor most was not the time savings on takeoff. It was the improvement in proposal quality. The AI-assisted proposals were more organized, more specific, and used more professional language than what they had been producing under time pressure. Two clients specifically mentioned that their proposal was the clearest and most detailed they received.

Story Two: The Client Communication Problem

Profile: A residential remodeling company with 12 employees, annual revenue approximately $3.5 million.

This contractor had a reputation problem that she knew was real but did not know how to fix without adding administrative staff she could not afford. Her clients were doing good work and producing projects that looked great. But clients were calling constantly during projects, sometimes multiple times per week, asking what was happening.

She implemented a simple weekly update system using ChatGPT. Every Friday morning, she spent twenty to twenty-five minutes dictating status notes for each of her seven active projects, putting them through the AI tool, reviewing the outputs, and sending them. The emails were specific, they included what was happening, what was coming, and whether the client needed to decide anything.

Within three weeks, the unsolicited calls dropped noticeably. By the end of the second month, she was tracking about a 65 percent reduction in mid-project client calls. The calls she was getting were from clients who had questions about their update or wanted to follow up on something specific, which was exactly the kind of communication she wanted to be having.

But the result she did not expect was in her end-of-project reviews. Client satisfaction scores on the metric of communication improved significantly. Several clients specifically mentioned feeling well-informed throughout the project. Referrals in the twelve months following the implementation were up substantially compared to the prior year.

Story Three: The Scope Gap That Did Not Happen

Profile: A mid-size GC doing mostly commercial construction, annual revenue approximately $12 million.

This contractor had learned an expensive lesson a few years before. A scope dispute with a mechanical sub on a medical office project had cost him nearly $80,000, split between a change order he had to absorb, direct legal costs, and the time his project manager spent managing the dispute instead of running projects.

After that experience, he started running scope gap analyses using Claude on every major project before finalizing his sub agreements. He would paste his project scope and each sub's bid side by side and ask the AI to identify any items in the project scope that were not covered by the sub's proposal.

On his next large project, a 12,000 square foot office building, this exercise flagged two gaps in the electrical scope: temporary power and site lighting during construction were in the project spec but not clearly addressed in any sub's bid. He went back to his electrical sub for clarification before executing the subcontract. The sub agreed to include both items with a modest price adjustment.

He estimates this finding saved him between $15,000 and $25,000 in potential change orders or dispute costs on that single project. The analysis took him about fifteen minutes.

Story Four: The Documentation That Won the Argument

Profile: A small residential GC, annual revenue approximately $1.8 million, primarily kitchens and bathrooms.

This contractor started using the five-minute daily log approach after reading about it in an industry newsletter. He used the voice dictation on his phone to note site conditions, crew composition, and key decisions at the end of each day, then sent the transcription to Claude to format as a professional log.

Six months later, a client disputed the cause of a crack that appeared in a bathroom tile installation after the project was complete. The client claimed the crack was the result of improper installation. The contractor believed it was caused by the client's own activities after project completion and by movement in the subfloor that he had documented as a concern during installation.

He pulled his daily logs from the project. The logs documented that the subfloor had exhibited movement that exceeded tolerances, that he had raised this with the client during installation and offered two options, that the client had chosen to proceed without the additional reinforcement work, and that this conversation and the client's decision were documented on a specific date.

The dispute was resolved in his favor. The client had no documentation of their own to counter the contemporaneous logs. The total time he had invested in documentation on that project was less than an hour and a half spread across the six-week job.

Common Themes Across the Stories

Looking across these four stories and the broader patterns of AI adoption among contractors, a few consistent themes emerge.

The Time Investment Is Smaller Than Expected - Almost every contractor who has adopted AI tools consistently reports that the learning curve was shorter than they anticipated. Building basic fluency with a general AI tool takes about a week of daily use. Building a functional prompt library for your most common workflows takes about a month.

The Benefits Are Often Indirect - The direct time savings on estimating and documentation are measurable and significant. But many of the contractors who report the highest satisfaction with AI adoption are getting benefits that are harder to quantify: improved proposal quality, more consistent client communication, fewer disputes, and a stronger professional reputation.

The Biggest Gains Come from Consistency - The contractors who get the most from AI tools are the ones who use them consistently, every week, across multiple workflows.

Team Adoption Multiplies the Value - When AI tools are used by one person in a business, the benefits are real but limited. When a team of five or ten people is all using AI for their respective workflows, the aggregate time savings are substantial. Getting your team to adopt consistently takes clear workflows, shared prompt libraries, and an environment where people feel safe experimenting with new tools.

CHAPTER 15

What Comes Next

This final chapter is not about predictions. Nobody knows exactly how AI will develop over the next five years, and anyone who claims certainty about the timeline or the specific form of future changes is guessing. What this chapter does is give you a grounded view of the directions that seem most likely, based on what is happening in the construction industry today, so that you can position your business thoughtfully rather than reactively.

The underlying theme is this: the contractors who will thrive in the next decade are the ones building systems and capabilities today, not the ones who are reacting to each new development as it arrives. AI is one of the most important tools available for building those systems.

Where AI in Construction Is Heading

From Tool to Teammate

The AI tools available today are largely reactive. You prompt them, they respond. The direction the industry is moving, and moving fairly quickly, is toward AI that is more proactive: systems that monitor your project data, surface alerts when something looks off, suggest actions before you ask, and coordinate across multiple systems without requiring you to initiate every interaction.

Procore's acquisition of Datagrid in early 2025 is an example of this direction. Datagrid describes its technology as agentic AI, meaning AI that can take sequences of actions on its own to accomplish a goal, rather than just responding to individual prompts. For a GC, this eventually means AI that monitors your schedule and alerts you when a trade dependency is at risk before it becomes a delay, not after.

This evolution will happen gradually. The tools of 2026 and 2027 will be more capable than today's tools, but not radically different in how you interact with them. Getting comfortable with AI now means you will be positioned to adopt more powerful tools as they arrive, rather than starting from scratch.

Specialized Models Trained on Construction Data

One of the limitations of current general-purpose AI tools is that they are trained on general text data, not specifically on construction projects. The next generation of construction AI tools will increasingly be trained on construction-specific data and allow you to fine-tune them on your own historical project data. When an estimating model can learn from your actual completed projects, your actual costs, and your actual market conditions, the accuracy of its outputs will improve significantly over what is possible today.

Computer Vision on the Jobsite

Computer vision, the ability of AI to analyze images and video, is being applied to construction sites in ways that are just beginning to reach practical usability for contractors of all sizes. Tools like OpenSpace use 360-degree cameras and AI to track site progress against plans automatically. Safety monitoring systems are beginning to use camera feeds to detect missing PPE, restricted zone breaches, and unsafe behaviors in real time.

The direction this is heading is toward AI systems that can monitor a jobsite continuously, provide real-time alerts about safety issues, automatically generate progress documentation, and flag quality deviations as they occur rather than at inspection.

Sector Opportunities Worth Positioning For

Data Center Construction

The demand for data centers driven by AI, cloud computing, and digital infrastructure has created a significant pipeline of construction work. Major technology companies are spending hundreds of billions of dollars on data center expansion, and the timeline extends well beyond the next few years. GCs who develop experience with this project type early will be in a strong position as the pipeline continues to grow.

Renewable Energy Infrastructure

Solar, battery storage, and grid infrastructure projects represent another major pipeline of construction work. The energy transition is driving billions in capital investment toward projects that require civil, structural, electrical, and specialty contractors. Many of these projects are in suburban and rural markets where smaller and mid-size GCs can be competitive.

Infrastructure Modernization

Federally funded infrastructure investment continues to create demand for transportation, water, and utility construction across the country. GCs with strong documentation practices, professional proposal capabilities, and the ability to navigate public procurement requirements are well positioned for this work.

Building a Business That Lasts

AI adoption is not ultimately about technology. It is about building a more resilient, more professional, and more scalable construction business. The practical outcomes of the workflows covered in this book are faster estimates, better proposals, clearer scopes, more consistent client communication, stronger documentation, and more organized sub management.

The goal for most owner-operator GCs is not necessarily to build a large company. It is to build a company that gives you the financial stability you need, delivers quality work you are proud of, and does not require you to be personally involved in every detail of every project every day. That kind of business is built through systems, not through working harder.

AI is one of the most accessible and affordable tools available for building those systems. The cost of the tools in this book is measured in hundreds of dollars per month, not tens of thousands. The time investment to build fluency is measured in hours and weeks, not months and years.

THE LONG VIEW
Better documentation protects you in disputes.
Better communication keeps clients satisfied.
Better estimating wins more work at the right margin.
Better systems mean you do not have to be personally present in every decision.
AI helps with all of it. Start this week.

A Final Word

If you have read this far, you now have more practical knowledge about using AI in your construction business than the vast majority of your competitors. That matters!

But knowledge without action is just reading. The most important thing you can do right now is pick one thing from this book and try it today. Not next week when things slow down. Not after the current project wraps up. Today.

Open ChatGPT or Claude on your phone. Pull a prompt from Chapter 10. Type in a description of your project. See what comes back. The first result will probably surprise you, either because it is better than you expected or because you immediately see how to improve it with your own knowledge. Both reactions are progress.

This industry has always been built on people who figure things out. Contractors are some of the most resourceful, adaptable problem solvers there are. A general contractor is, by definition, someone who can hold a complex project in their head, coordinate dozens of different parties, manage competing priorities, and still make good decisions under pressure. That is exactly the kind of person who can use AI effectively.

The tools in this book are starting points. Use them, refine them, build on them. Share the ones that work with your team. Add the prompts that fit your specific project types. Over time, your AI workflows will be more tailored to your business than anything described in any general-purpose guide.

The contractors who thrive in the next decade will not necessarily be the ones with the best field crews or the most experience, though those things still matter enormously. They will be the ones who build the most efficient, most professional, and most adaptable operations. AI is one of the most powerful tools available for doing exactly that.

Go build something.

★ ★ ★ ★ ★

Thank You for Reading!

We hope you enjoyed this book and found value in its content. Please consider leaving an honest review. It takes just a minute. Reviews are so important to independent publishers.

LEAVE A REVIEW

To see more titles from this publisher, visit:

arrowjaypublishing.com

Made in the USA
Monee, IL
13 September 2026

62980081R00066